Audel™

Machine Shop Basics
All New 5th Edition

Rex Miller
Mark Richard Miller

WILEY

Wiley Publishing, Inc.

Vice President and Executive Group Publisher: Richard Swadley
Vice President and Executive Publisher: Robert Ipsen
Vice President and Publisher: Joseph B. Wikert
Executive Editorial Director: Mary Bednarek
Editorial Manager: Kathryn A. Malm
Executive Editor: Carol A. Long
Senior Production Manager: Fred Bernardi
Development Editor: Kevin Shafer
Production Editor: Vincent Kunkemueller
Text Design & Composition: TechBooks

Contents

Acknowledgments

A number of companies have been responsible for furnishing illustrative materials and procedures used in this book. At this time, the authors and publisher would like to thank them for their contributions. Some of the drawings and photographs have been furnished by the authors. Any illustration furnished by a company is duly noted in the caption.

No book can be written and illustrated and published by the authors without the assistance of others. Some of those who have worked on this book through the many editions were Perry Black, Bob Wellboarn, John Obst, Jack Brueckman, Thomas Morrisey, and Ellsworth Russell. We would like to take this opportunity to thank each and every one of them for their valuable contributions.

The authors would like to thank everyone involved for his or her contributions. Some of the firms that supplied technical information and illustrations are listed below:

American Abrasive Company

American Twist Drill Company

Black & Decker

Brown & Sharpe Manufacturing Company

Cincinnati Milicron, Inc.

Disston Company

Do All Company

Greenfield Tap & Die Company

Hamilton Watch Company

Heald Machine Company

L.S. Starrett Company

Lufkin Tool Company

Monarch-Cortland

Millers Falls Company

Morse Twist Drill & Machine Company

National Twist Drill Company

National Association of Fire Equipment Distributors

Newage, Inc.

Nicholson File Company

Norton Company
Reliance Electric Company
Ridge Tool Company
Shore Instrument Company
Simonds Saw and Steel Company
South Bend Lathe Company
Stanley Tools Company
The Heald Machine Company
Tinius Olsen Testing Machine Company
Westinghouse Electric Corp.
Wilson Mechanical Instrument Company
Wilson Tool Manufacturing

About the Authors

Rex Miller was a Professor of Industrial Technology at The State University of New York, College at Buffalo for over 35 years. He has taught on the technical school, high school, and college level for well over 40 years. He is the author or co-author of over 100 textbooks ranging from electronics to carpentry and sheet metal work. He has contributed more than 50 magazine articles over the years to technical publications. He is also the author of seven civil war regimental histories.

Mark Richard Miller finished his BS degree in New York and moved on to Ball State University, where he obtained a master's degree and went to work in San Antonio. He taught in high school and went to graduate school in College Station, Texas, finishing a doctorate degree. He took a position at Texas A&M University in Kingsville, Texas, where he now teaches in the Industrial Technology Department as a Professor and Department Chairman. He has co-authored seven books and contributed many articles to technical magazines. His hobbies include refinishing a 1970 Plymouth Super Bird and a 1971 Roadrunner. He is also interested in playing guitar, an interest he pursued while in college as the lead guitarist of a band called The Rude Boys.

Introduction

The purpose of this book is to provide a better understanding of the fundamental principles of working with metals in many forms, but with emphasis on machining—utilizing both manually operated and automated machines. The beginner and the more advanced machinist alike may benefit from studying the procedures and materials shown in these pages.

One of the chief objectives has been to make the book clear and understandable to both students and workers. The illustrations have been selected to present the how-to-do-it phase of many of the machine shop operations. The material presented here should be helpful to the machine shop instructor, as well as to the individual student or worker who desires to improve himself or herself in this trade.

The proper use of machines and the safety rules for using them have been stressed throughout the book. Basic principles of setting the cutting tools and cutters are dealt with thoroughly, and recommended methods of mounting the work in the machines are profusely illustrated. The role of numerically controlled machines is covered in detail with emphasis on the various types of machine shop operations that can be performed by them.

This book has been developed to aid you in taking advantage of the trend toward vocational training of young adults. An individual who is ambitious enough to want to perfect himself or herself in the trade will find time to do so. Or, an apprentice working under close supervision on the job can also benefit from using this material.

Chapter 1

Benchwork

The term *benchwork* relates to work performed by the mechanic at the machinist's bench with hand tools rather than machine tools. It should be understood that the terms *benchwork* and *visework* mean the same thing; the latter, strictly speaking, is the correct term, as in most cases the work is held by the vise, while the bench simply provides an anchorage for the vise and a place for the tools. However, these terms are used almost equally. Today, work at the bench is not performed as much as formerly; the tendency, with the exception of scraping, is to do more and more benchwork with machines.

Operations that can be performed at the bench may be classed as follows:

- Chipping
- Sawing
- Filing
- Scraping

The Bench and Bench Tools

The prime requirements for a machinist's bench are that it should be strong, rigid, and of the proper width and height that the work can be performed conveniently. Correct height is important, and this will depend on the vise type used, that is, how far its jaws project above the bench. The location of the bench is important. It should be placed where there is plenty of light.

A great variety of tools is not necessary for benchwork. They may be divided into a few general classes:

- Vises
- Hammers
- Chisels
- Hacksaws
- Files
- Scrapers

Vises

By definition, a vise is a clamping device, usually consisting of two jaws that close with a screw or a lever, that is commonly attachable to a workbench; it is used for holding a piece of work firmly. There is a great variety of vises on the market, and they may be classed as follows:

- Blacksmith
- Machinist (plain, self-adjusting, quick-acting, or swivel)
- Combination
- Pipe

The machinist's vise shown in Figure 1-1 is usually provided on machine shop workbenches. Several types are provided; some of their features are parallelism, swivel action, and quick-acting jaws. These vises will withstand terrific abuse and are well adapted for a heavy and rough class of work.

Figure 1-1 Machinist's vise. *(Courtesy Ridge Tool Company.)*

The combination vise shown in Figure 1-2 is well adapted for round stock and pipe. A regular pipe fitter's vise is shown in Figure 1-3. Vise jaws have faces covered with cross cuts in order to grip the work more firmly. It is evident that a piece of finished work held in such a manner would be seriously marred. This trouble may be avoided by using false jaws of brass or Babbitt metal, or by fastening leather or paper directly to the steel jaws.

Figure 1-2 Combination vise. The inner teeth are for holding either pipe or round stock. *(Courtesy Wilton Tool Mfg. Co.)*

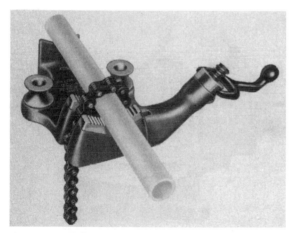

Figure 1-3 Pipe vise. *(Courtesy Ridge Tool Company.)*

Hammers

Hammers find frequent use in benchwork. Machinist's hammers may be classed with respect to the peen as follows:

- Ball peen
- Straight peen
- Cross peen

By definition, *peening* is the operation of hammering metal to indent or compress it in order to expand or stretch that portion of the metal adjacent to the indentation. These hammers are shown in Figure 1-4. The ball-peen hammer (Figure 1-4A), with its spherical end, is generally used for peening or riveting operations. For certain classes of work, the straight indentations of either the straight- or cross-peen hammers (Figure 1-4B and 1-4C) are preferable. A shaft or bar may be straightened by peening on the concave side.

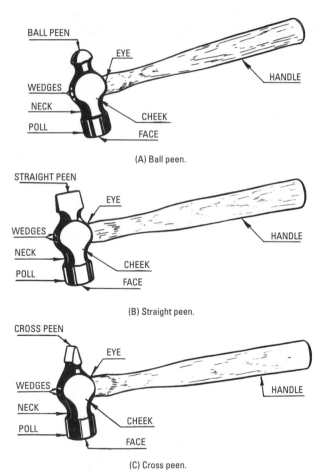

(A) Ball peen.

(B) Straight peen.

(C) Cross peen.

Figure I-4 Machinist's hammers.

Chisels

The cold chisel is the simplest form of metal cutting tool. By definition, a chipping chisel is a hand tool made of heat-treated steel, with the cutting end shaped variously, for chipping metal when it is struck by a hammer.

The various types of chipping chisels are as follows:

- Flat
- Cape
- Diamond-point
- Round-nose

One of the first operations that a student or apprentice must learn in becoming a machinist is how to chip metal. This involves learning how to hold the chisel and how to use the hammer.

Flat Chisel

The work is placed firmly in the vise with the chisel held in the left hand. The chisel must be held firmly at the proper angle (Figure 1-5) to the work. The lower face of the chisel cutting edge acts as a guide, while the wedging action of the metal being chipped tends to guide the chisel in a straight line. The cutting face is the guide to hold the chisel at the correct angle, as shown in Figure 1-6. The cutting edge of the chisel is ground at an included angle of 60° to 70° (Figure 1-6A).

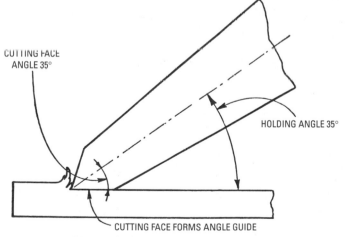

CUTTING FACE ANGLE 35°

HOLDING ANGLE 35°

CUTTING FACE FORMS ANGLE GUIDE

Figure 1-5 Detail of cutting end of cold chisel. Note the angle of application and the angle guide.

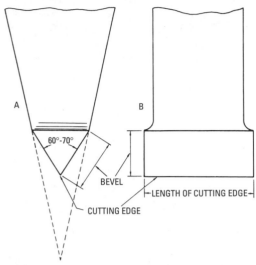

A 60°-70° BEVEL CUTTING EDGE

B |—LENGTH OF CUTTING EDGE—|

Figure 1-6 Cutting end of cold chisel showing bevel angle (A) and length of cutting edge (B).

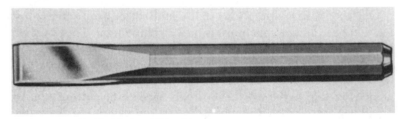

Figure 1-7 Cold chisel used in chipping operations. *(Courtesy Millers Falls Company.)*

The flat chisel is used for surfaces having less width than the castings and for all general chipping operations (Figure 1-7). The cutting edge is generally about one-eighth of an inch wider than the stock from which it is forged.

The beginning machinist learns to vary the chipping angle more or less as demanded by the nature of the work. The first exercise in chipping is usually a broad surface on which both the cold chisel and the cape chisel are used. First, grooves are cut in the piece to be chipped with the cape shield (Figure 1-8), and the raised portions are removed with a flat chisel (Figure 1-9).

In chipping, the worker should always chip toward the stationary jaw of the vise because its resistance to the blows of the hammer

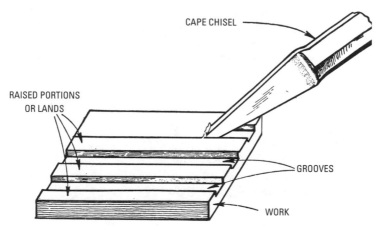

Figure 1-8 Cape chisel used to cut grooves.

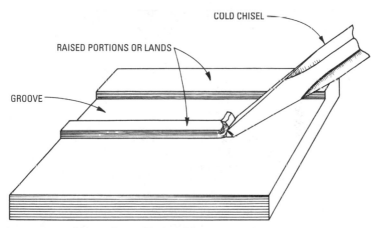

Figure 1-9 Using the cold chisel to remove *lands* in chipping a broad flat surface.

is greater. Start with a light chip, and watch only the cutting edge of the chisel. Chamfer the front and back edges of the work to avoid risk of breaking off the stock below the chipping line and to facilitate starting the chisel.

Use a 1-to 1¾-lb hammer for ordinary chipping work. Grasp the hammer near the end of the handle, with the fingers around the handle and the thumb projecting on top toward the striking end.

The chisel should be held firmly with the second and third fingers, and the little finger should be used to guide the chisel as required. The first finger and the thumb should be left slack; they are then in a state of rest, with the muscles relaxed. The fingers are less liable to become injured if struck with the hammer when relaxed than if struck when they were closed rigidly around the chisel. Reset the chisel to its proper position after each blow.

Cape Chisel

A cape chisel (Figure 1-10) is used to facilitate work in removing considerable metal from a flat surface, or to break up surfaces too wide to chip with a cold chisel alone. It is also used, along with other chisels, to cut keyways and channels.

Figure 1-10 Cape chisel. *(Courtesy Millers Falls Company.)*

The cutting edge of the cape chisel is usually an eighth of an inch narrower than the shank. It is thin enough just behind the cutting edge to avoid binding in the slot. It is somewhat thicker in the plane at a right angle to the cutting edge.

Diamond-Point Chisel

Although the word "point" is universally used in place of "end," the term is a misnomer. The diamond end is obtained by drawing out the end of the stock and grinding the end at an angle less than 90° with the axis of the chisel, leaving a diamond-shaped point (Figure 1-11).

The diamond-point chisel (Figure 1-12) is used by diemakers for corner chipping, for correcting errors made while drilling holes, and for cutting holes in steel plates. By cutting a groove with this tool, following the shapes to be cut in the piece is much easier. The edges of holes made this way will be beveled, but they can be chipped square after the piece is removed.

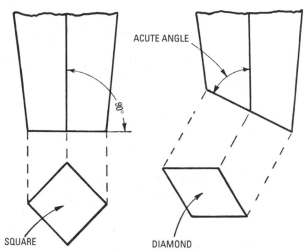

Figure 1-11 Detail of the cutting end of square and diamond-point chisels.

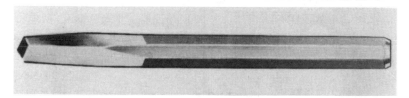

Figure 1-12 Diamond-point chisel. *(Courtesy Millers Falls Company.)*

Round-Nose Chisel

The round-nose chisel is sometimes called a *round-nose cape chisel* (Figure 1-13). The nose itself is cylindrical in section at the cutting end with tangential sides intersecting at the extremity. The tool is ground at an angle of 60° with its axis.

These chisels are called center chisels when they are used to "draw" the starting of drilling holes in order to bring them into concentricity with the drilling circles. They are also used on large round-bottomed channels and for cutting channels such as oil grooves.

The stock generally used for all the aforementioned forms of chisels is octagonal and of a good grade of tool steel, carefully forged, hardened, and tempered.

Figure 1-13   Round-nose cape chisel or round-nose chisel. *(Courtesy Millers Falls Company.)*

Hacksaws

The sawing of metal is one of the most common benchwork operations. Hand hacksaws are available with either a fixed frame or an adjustable frame. The adjustable frame hacksaw (Figure 1-14) can be changed to hold 8-inch, 10-inch, and 12-inch blades. Most blades are ½-inch wide and ¼-inch thick.

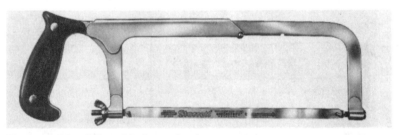

Figure 1-14   Adjustable frame hacksaw. *(Courtesy L. S. Starrett Company.)*

The workpiece must be held securely in a vise. The workpiece should be sawed near the vise jaws to prevent chattering. To hold nonrectangular-shaped pieces (Figure 1-15), clamp the work to allow as many teeth as possible to be in contact with the surface of the workpiece. Polished work surfaces should be protected from the steel vise jaws by covering them with soft metal jaw caps.

The type of metal to be sawed should determine the blade pitch (number of teeth per inch). Standard pitches are 14, 18, 24, and 32 teeth per linear inch. The number of teeth per inch on a blade is important because at least two teeth should be in contact with the work at all times (Figure 1-16).

To start a hacksaw cut, it is a good practice to guide the blade until the cut is well established. To start an accurate cut, use the thumb (Figure 1-17) as a guide and saw slowly with short strokes.

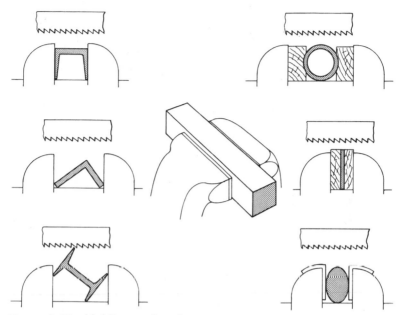

Figure 1-15   Holding work to be cut. *(Courtesy Disston, Inc.)*

As the cut deepens, grip the front end of the frame firmly and take a full-length stroke.

When sawing, stand facing the work with one foot in front of the other and approximately 12 inches apart, as shown in Figure 1-18. Pressure should be applied on the forward stroke and released on the return stroke because the blade cuts only on the forward stroke. Do not permit the teeth to slip over the metal as this dulls the teeth and may cause blade breakage. Once the *kerf* (the slot made by the blade) is established, the hacksaw should be moved at about 40 strokes per minute.

Files

Filing is a difficult operation for the beginner because it depends on the motion of the hands, without a means of guiding the tool, to move over the work in the correct direction. Proficiency is obtained by practice only when the proper methods are followed.

How to File

The correct position and method of holding the file are important. The work should be at the proper height—about level with the

USE 14 TEETH
For Softer Larger Sections

For cutting material 1" or thicker in sections of cast iron, machine steel, brass, copper, aluminum, bronze, slate.

USE 24 TEETH
For Angle Iron, Brass, Copper, Iron Pipe, Etc.

For cutting material 1/8" to 1/4" in sections of iron, steel, brass and copper tubing, wrought iron pipe, drill rod, conduit, light structural shapes, metal trim.

INCORRECT

COARSE TEETH
STRADDLES WORK
STRIPS TEETH

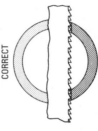

CORRECT

THREE TEETH OR
MORE ON SECTION

USE 32 TEETH

For Conduit and Other Thin Tubing, Sheet Metal Work
For cutting material similar to recommendations
for 24 tooth blades for 1/8" and thinner.

CORRECT

PLENTY OF
CHIP CLEARANCE

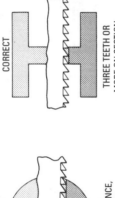

INCORRECT

FINE TEETH,
NO CHIP CLEARANCE,
TEETH CLOGGED

USE 18 TEETH
For General Use

For cutting materials 1/4" to 1" in sections of annealed tool steel, high speed steel, rail, bronze, aluminum, light structural shapes, copper.

INCORRECT

PITCH TOO COARSE
TEETH STRADDLE WORK

CORRECT

THREE OR MORE TEETH
IN EACH WALL SECTION

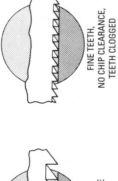

INCORRECT

TEETH TOO FINE,
NO CHIP CLEARANCE,
TEETH CLOGGED

CORRECT

AMPLE CHIP
CLEARANCE

Figure 1-16 Hacksaw blade selection for various cutting operations. (Courtesy Disston, Inc.)

12

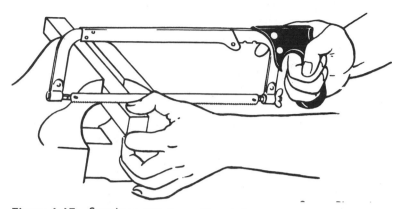

Figure 1-17 Starting a cut. *(Courtesy Disston, Inc.)*

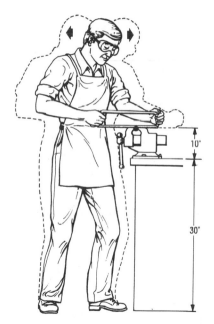

10"

30"

Figure 1-18 Proper stance for cutting. *(Courtesy Disston, Inc.)*

elbows on light work, and a little lower on heavy work (Figure 1-19). The feet should be about eight inches apart and at right angles to each other, the left foot being parallel with the file. Hold the file handle with the right hand-thumb on top and fingers below the handle.

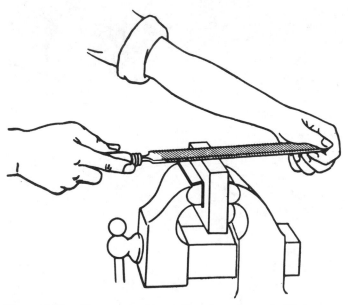

Figure 1-19   Correct position of hands and arms in filing. *(Courtesy Disston, Inc.)*

When filing, pressure should be exerted on the forward stroke only, because the teeth or cutting edges are pointed toward the end of the file. Pressure on the return stroke produces no cutting action, but tends only to dull the teeth. Figure 1-20 shows an incorrect position of the body when filing.

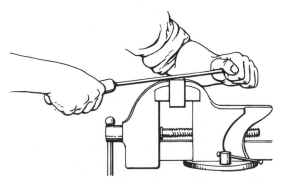

Figure 1-20   Incorrect position of body when filing.
(Courtesy Nicholson File Company.)

Drawfiling

When the file is grasped by the ends and moved sidewise across the work, the action is known as drawfiling (Figure 1-21). This produces a smooth finish on narrow surfaces and edges and is used on turned work to remove any tool marks. Drawfiling is light filing—used to produce a smooth surface (Figure 1-22). A second-cut or smooth file should be used; a single-cut file is better than a double-cut file because the single-cut is less likely to scratch the surface of the work.

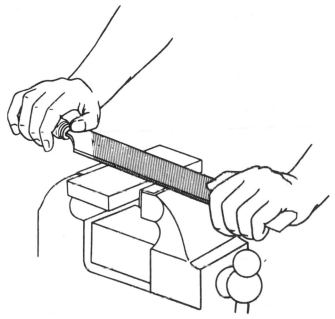

Figure 1-21 Drawfiling for producing a smooth surface. *(Courtesy Disston, Inc.)*

For most filing operations, begin with a coarse file and continue using successively finer grades of file, finishing with a smooth or dead-smooth file, according to the degree of finish desired (Figure 1-23).

Particles of metal, or pins, often remain in the teeth of the file, and they either reduce its cutting qualities or scratch the work. These particles can be removed by using either a stiff brush (Figure 1-24A) or a file card (Figure 1-24B) frequently for cleaning them from the file.

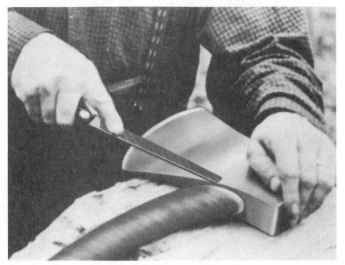

Figure 1-22 Using one hand to do light filing. *(Courtesy Nicholson File Company.)*

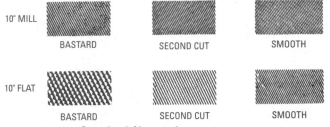

10" MILL   BASTARD   SECOND CUT   SMOOTH

10" FLAT   BASTARD   SECOND CUT   SMOOTH

Figure 1-23 Standard file tooth cuts. *(Courtesy Simonds Saw & Steel Company.)*

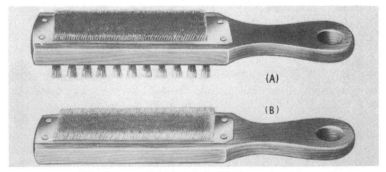

Figure 1-24 File cleaners: (A) file brush, (B) file card. *(Courtesy Nicholson File Company.)*

Cast iron should not be allowed to become greasy, as the file tends to slide without cutting into the metal. However, frequent *pinning* (clogging of the teeth with small slivers of metal) can be prevented by the use of oil when filing steel.

File Characteristics

A file differs from a chisel in that it has a large number of cutting points instead of a single cutting edge, and the file is driven by hand, rather than by a hammer. When a file is applied to a metal surface with a reciprocating motion, the teeth act as small chisels, each removing small chips.

Files have three distinguishing characteristics (Figure 1-25):

- *Length*—Always measured from the heel to the point, the tang not being included
- *Kind*—The shape or style
- *Cut*—Both the character and the relative degrees of coarseness of the teeth

Length

File lengths vary from 3 inches to 20 inches. Most machinist's files are from 4 to 6 inches in length (Figure 1-25A).

Kind

Many kinds of files are manufactured for many different purposes. Shapes of files in common use are mill, flat, hand, square, three-square, half-round, and round files (Figure 1-25B).

Cut

The teeth on a file are shaped to form a cutting edge similar to that of a tool bit, and they have both rake and a clearance angle. Four types of cuts are shown in Figure 1-25C. *Single-cut* files are made with a single set of teeth cut at an angle of 65° to 85°. They are usually used with light pressure to produce a smooth finish on a surface or to produce a keen edge on a knife or other cutting implement. *Double-cut* files are made with two sets of teeth that cross each other. One set is cut at approximately 45° and the other set at 70° to 80°. On a *rasp-cut* file, each tooth is short and is raised out of the surface by means of a punch. A *vixen-cut* file (or *curved-tooth* file) has a series of parallel, curved teeth, each extending across the file face. Most files for hand filing are from 10 to 14 inches long and have a pointed tang on one end on which wood or metal handles can be fitted for easy grasping.

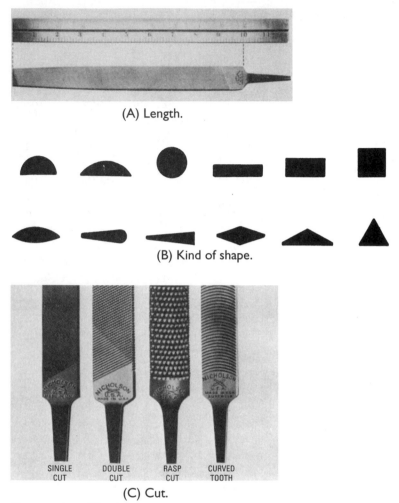

(A) Length.

(B) Kind of shape.

SINGLE CUT   DOUBLE CUT   RASP CUT   CURVED TOOTH

(C) Cut.

Figure 1-25   File characteristics. *(Courtesy Nicholson File Company.)*

Machinist's files (Figure 1-26) are used throughout the industry wherever metal must be removed rapidly and finish is of secondary importance. They include flat, hand, round, half-round, square, pillar, three-square, warding, knife, and several less commonly known kinds of files. Most machinist's files are double-cut (Figure 1-27).

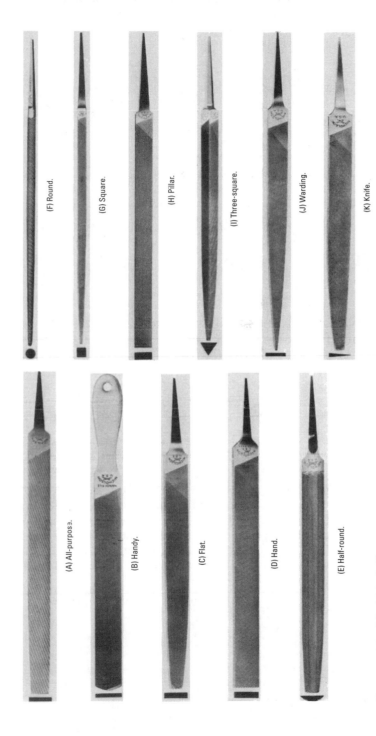

(F) Round.

(G) Square.

(H) Pillar.

(I) Three-square.

(J) Warding.

(K) Knife.

(A) All-purpose.

(B) Handy.

(C) Flat.

(D) Hand.

(E) Half-round.

Figure 1-26 Machinist's files. *(Courtesy Nicholson File Company,*

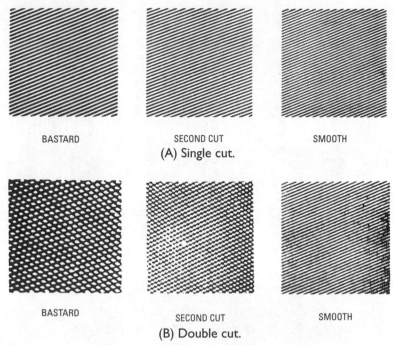

BASTARD   SECOND CUT   SMOOTH
         (A) Single cut.

BASTARD   SECOND CUT   SMOOTH
         (B) Double cut.

Figure 1-27   Single-cut and double-cut files. Each type has its own application. *(Courtesy Simonds Saw & Steel Company.)*

The cut (coarseness) of small files is usually designated by numbers as 00, 0, 1, 2, 3, 4, 5, 6, 7, and 8. The coarsest cut is 00, and 8 is the finest cut (Figure 1-28). The cut or coarseness in larger files is designated as rough, coarse, bastard, second-cut, smooth, and dead-smooth. These designations are relative and depend on the length of a file. A 14-inch bastard file is much coarser than a 6-inch bastard file (Figure 1-29.)

Scrapers

Scraping is the operation of correcting the irregularities of machined surfaces by means of scrapers (Figure 1-30) so that the finished surface is a plane surface. Although it is impossible to produce a true plane surface, scrapers are used to approach a plane surface, or for truing up a plane surface. Scrapers are also used for truing up circular surfaces such as bearings.

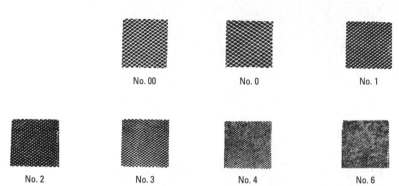

No. 00   No. 0   No. 1

No. 2   No. 3   No. 4   No. 6

Figure 1-28   Small files are designated by numbers from 00 to 8.
(Courtesy Nicholson File Company).

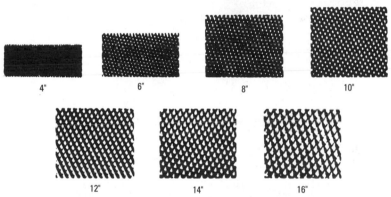

4"   6"   8"   10"

12"   14"   16"

Figure 1-29   The coarseness varies for flat bastard files used by machinists.
Length of the file determines coarseness. *(Courtesy Nicholson File Company.)*

How to Use a Scraper

In scraping operations (Figure 1-31), a surface plate is used to indicate irregularities or high spots. Any dust or grit should be wiped off the surface, and any burrs on the metal should be removed with a very fine file.

After thoroughly cleaning the surface plate, coat it with a marking material and rub the work over the surface plate a few times. High spots on the work will be indicated by the marking material that has been rubbed off. These high spots are removed by scraping. Continuing the process will bring up more high spots. After

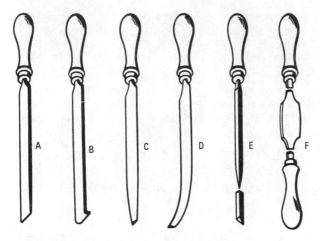

Figure 1-30 Typical scrapers: (A) flat or straight, (B) hook, (C) straight half-round, (D) curved half-round, (E) three-cornered or triangular, (F) double-handle.

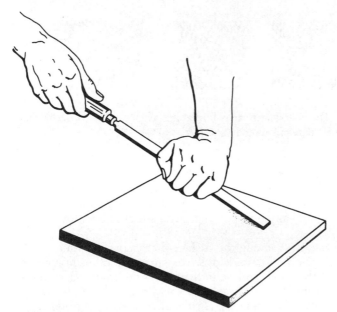

Figure 1-31 Correct method of holding a scraper.

repeated scraping and testing with the surface plate, the entire work surface will be covered with marking material, which indicates that the work is finished.

The correct use of the scraper is important. When a flat scraper is used, cutting is done on the forward stroke. Cutting is done on the return stroke when a hook scraper is used. Scraping requires a delicate touch. Less metal is removed by a scraper than by a file. The cutting operation, as done with scrapers, should be perfectly smooth and free from scratches.

The cutting edge of a scraper should be $3/32$ of an inch thick and $1\frac{1}{2}$ inches wide. The scraper should be ground on a grinding wheel and carefully honed on an oilstone. Scrapers are sometimes made from discarded files.

Scraper Classifications

Various forms of scrapers are used. The nature of the scraping operation determines the selection of the scraper. Scrapers may be classified as follows:

- Flat
- Hook (right-hand or left-hand)
- Half-round
- Triangular or three-cornered
- Two-handled
- Bearing

Scraping is also performed on round or curved surfaces, such as bearings. When an engine's main bearing has been trued up by scraping, the shaft will contact the bearing over its entire surface instead of making contact only at the high spots. Consequently, the bearing surface is presented, and the pressure is distributed over the entire bearing instead of being concentrated on the high spots.

Summary

Benchwork relates to work performed by the mechanic at the machinist's bench with hand tools rather than machine tools. The terms benchwork and visework mean the same thing. Visework, strictly speaking, is the correct term, as in most cases the work is held in a vise. Benchwork operations include chipping, sawing, filing, and scraping. Bench tools are the hammer, chisel, hacksaw, file, scraper, and vise.

The vise is a clamping device that has a couple of jaws that are used to hold a piece being worked on tightly in its grip. Vises are classified as blacksmith, machinist, combination, or pipe. The machinist vise is also classified as self-adjusting, quick-acting, plain, or swivel type. The jaws of the vise may be lined with brass or Babbitt metal or leather or paper to protect the piece being held rigid while the work is being done.

The machinist's hammer (ball-peen, straight-peen, or cross-peen) is suited only for the work it was designed to do. Ball-peen hammers are used for peening or riveting operations. But, for some types of straight work the straight peen is used.

Chisels are another of the hand tools that come in handy in metalwork and in the machine shop. Various types of chipping chisels are the flat, cape, diamond-point, and round-nose chisels. Chisels are used to chip metal. Holding the chisel correctly is very important in getting the job done.

The sawing of metal is one of the most important benchwork operations. Hand hacksaws are available with either a fixed frame or an adjustable frame. The work piece is held firmly in a vise while the work is being performed. It is very important that you use a hacksaw blade with a saw tooth fitted for the job. The hacksaw blade is made with 14, 18, 24, or 32 teeth per linear inch. The number of teeth is important because at least two teeth should be in contact with the work at all times.

Filing is a difficult operation for the beginner because it depends on the motion of the hands, without a means of guiding the tool, to move over the work in the correct direction. A lot of practice makes for a better filer. Small files are designated by numbers from 00 to 8. Drawfiling is the process of grasping the file by the ends and moving it sideways across the work. This produces a smooth finish on narrow surfaces and edges and is used on turned work to remove any tool marks.

Scraping is the operation of correcting the irregularities of machined surfaces by means of scrapers. It is very important to use the scraper correctly. Various forms of scrapers can be used by someone with a delicate touch to make a surface perfectly smooth. Scrapers are classified as flat, hook (right-hand, left-hand), half-round, triangular, two-handled, and bearing.

Review Questions

1. Name any five of the ten most popular machinist's files.

2. How many various forms of scrapers are used? Name them.

3. Name the three types of peen hammers and the four types of chisels.

4. Name the four types of bench vises.

5. Name five important bench tools.

6. Name the four hacksaw blade pitches.

7. How many hacksaw blade teeth should be in contact with the work piece?

8. What is the meaning of benchwork?

9. What are four operations that can be performed at the bench?

10. Name four types of chipping chisels.

11. What is drawfiling?

12. How does a file differ from a chisel?

13. What are the three distinguishing characteristics of files?

14. Why is scraping used on bearings?

15. Why does scraping take a delicate touch?

Chapter 2

Precision Measurement and Gaging

The worker in the machine shop uses many tools, instruments, and gages to produce accurate measurements. Precision measurements are generally written in decimals and are read in thousandths (0.001) and ten-thousandths (0.0001) of an inch.

Micrometer Calipers

The word *micrometer* indicates a precision instrument for small measurements, which are usually made by rotating a screw with a fine pitch. Micrometer calipers have a U-shaped frame with a hardened anvil at one end and an indicating thimble at the other end. The micrometer is the precision tool widely used for measurements in thousandths or even ten-thousandths of an inch (Figure 2-1). Micrometers are made in many styles and sizes for outside, inside, and depth measurements. The pitch of the screw thread on the spindle is $\frac{1}{40}$ inch (40 threads per inch). Therefore, one complete revolution of the thimble moves the spindle $\frac{1}{40}$ inch, or 0.025 inch, either toward or away from the anvil space. The longitudinal line on the sleeve is divided into 40 equal parts by vertical lines corresponding to the number of the threads on the spindle. Therefore, each vertical line designates $\frac{1}{40}$ inch, or 0.025 inch; every fourth line indicates hundreds of thousandths of an inch. The line marked "1" indicates 0.100 inch; the line marked "2" indicates 0.200 inch, and so on.

The beveled edge of the thimble is divided into 25 equal parts, each line representing 0.001 inch. To read the micrometer in thousandths of an inch, multiply the number of vertical divisions visible on the sleeve by 0.025 inch, and add the number of thousandths of an inch indicated by the line on the thimble, which coincides with the longitudinal line on the sleeve.

For example, the "1" line on the sleeve is visible, representing 0.100 inch. Three additional lines are visible, each representing 0.025 inch: 3 × 0.025 inch = 0.075 inch. Line "3" on the thimble coincides with the longitudinal line on the sleeve, each representing 0.001 inch: 3 × 0.001 inch = 0.003 inch. The micrometer reading is the total (0.100 + 0.075 + 0.003 = 0.178 inch, or 178 thousandths of an inch).

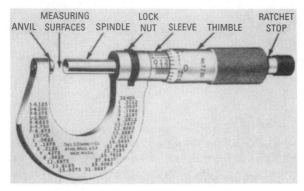

MEASURING SURFACES — ANVIL — SPINDLE — LOCK NUT — SLEEVE — THIMBLE — RATCHET STOP

(A) Parts of the micrometer.

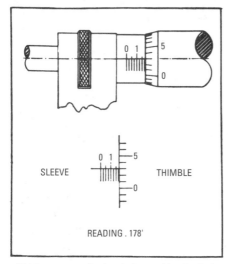

SLEEVE — THIMBLE

READING . 178'

(B) Micrometer graduations in thousandths of an inch.

Figure 2-1 Micrometer calipers graduated in thousandths of an inch.
(Courtesy L.S. Starrett Company.)

Micrometer calipers can be adjusted to compensate for wear. This is done by adjustment of the friction sleeve as follows (see Figure 2-2). Take up the wear of the screw and nut. Insert the spanner wrench in the slot of the adjusting nut and tighten just enough to eliminate play. Then, carefully bring the anvil and spindle together and insert the spanner wrench in the small slot of the sleeve. Turn the sleeve until the line on the sleeve coincides with the zero line on the thimble.

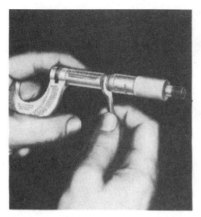
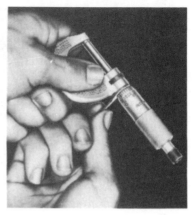

(A) Back off thimble, and tighten the adjusting nut to eliminate play in spindle nut.

(B) With anvil and spindle in contact, adjust sleeve, so that line on sleeve coincides with zero line on thimble.

Figure 2-2 Adjusting micrometer calipers. *(Courtesy L.S. Starrett Company.)*

When using a 1-inch micrometer for small work, hold the tool in one hand, turning the thimble with the thumb and forefinger (Figure 2-3A). This permits freedom for holding the work with the other hand. On some types of work (such as measuring over two flat surfaces), the micrometer is held in the left hand and the thumb and forefinger are used to turn the sleeve to adjust to the dimension. Round stock may be held in the left hand and the micrometer in the right hand. The thumb and forefinger are used to turn the sleeve until the correct setting is indicated by "feel" (Figure 2-3B). In larger work, or in stationary work, the frame should be held securely in one hand while the other hand turns the thimble (Figure 2-4).

Some mechanics change a micrometer setting quickly by holding the sleeve and swinging the frame around several times. This practice should be avoided because the centrifugal force generated by the whirling frame unduly wears the threads on the spindle, rendering the instrument inaccurate.

The inside micrometer is used for obtaining precision measurements of internal diameters of cylinders, holes, and so on, as shown in Figure 2-5.

Vernier Micrometer Calipers

Micrometers graduated in ten-thousandths of an inch are used in the same manner as the micrometers graduated in thousandths

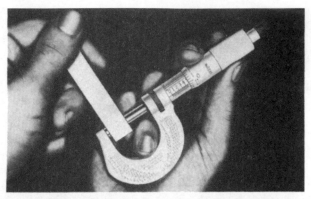

(A) Measuring a piece of die steel.

(B) Measuring tubular work.

Figure 2-3   Using micrometer calipers. *(Courtesy L.S. Starrett Company.)*

of an inch (Figure 2-6), except that an additional reading in ten-thousandths of an inch is obtained from a vernier and is added to the thousandth reading.

The vernier has ten divisions on the sleeve (Figure 2-6B), which occupy the same amount of space as nine divisions on the thimble. Therefore, the difference between the width of one of the ten spaces on the vernier and one of the nine spaces on the thimble is one-tenth of a division on the thimble, or one-tenth of one-thousandth, which is one ten-thousandth of an inch.

To read the ten-thousandths-of-an-inch micrometer, obtain the thousandths reading and note which of the lines on the vernier

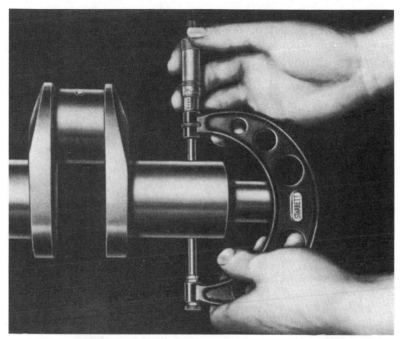

Figure 2-4 Measuring work with a large micrometer. Hold the frame securely in one hand at a convenient point. Turn the thimble with the other hand. *(Courtesy L.S. Starrett Company.)*

coincides with a line on the thimble. If line "1" on the vernier coincides, add one ten-thousandth; if line "2" coincides, add two ten-thousandths, and so on.

Vernier Calipers

By definition, *vernier calipers* have a graduated blade and an adjustable tongue (Figure 2-7). The blade has graduations and carries two crossheads, one of which is slightly adjustable by a nut, the other being movable along the blade. The crossheads are adapted to the measurement of interior diameters or sizes, and the other side is adapted to external measurements.

The bar of the calipers is graduated in fortieths, or 0.025, of an inch. Every fourth graduation is numbered to represent a tenth of an inch (Figure 2-8). The vernier scale is divided into 25 divisions, numbered 0, 5, 10, 20, and 25. The 25 divisions on the vernier scale occupy the same space as 24 divisions on the bar.

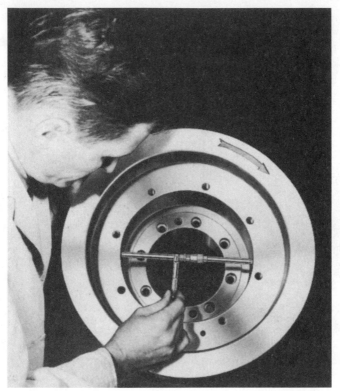

Figure 2-5 Measuring an inside diameter with an inside micrometer calipers. *(Courtesy L.S. Starrett Company.)*

Since one division on the bar is equal to 0.025 inch, 24 divisions equal 24 × 0.025 inch, or 0.600 inch, and 25 divisions on the vernier scale also equal 0.600 inch. Therefore, each division on the vernier is equal to ⅟₂₅ × 0.600 inch, or 0.024 inch. The difference between one bar division (0.025 inch) and one vernier division (0.024 inch) equals 0.025 inch less 0.024 inch, or 0.001 inch. If the zero line on the vernier coincides with the zero line on the bar, the line to the right of the zero on the vernier will differ from the line to the right of the zero on the bar by 0.001 inch, the second line by 0.002 inch, and so on. The difference increases by 0.001 inch for each division until the "25" on the vernier coincides with the "24" on the bar. In reading the calipers, note the number of inches, tenths (or 0.100 inch), and fortieths (or 0.025 inch) the zero line on the

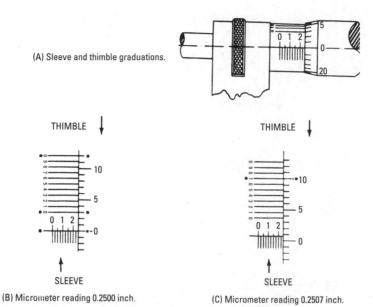

(A) Sleeve and thimble graduations.

THIMBLE ↓

THIMBLE ↓

SLEEVE

SLEEVE

(B) Micrometer reading 0.2500 inch.

(C) Micrometer reading 0.2507 inch.

Figure 2-6 A micrometer graduated in ten-thousandths of an inch.
(Courtesy L.S. Starrett Company.)

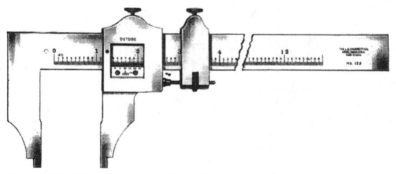

Figure 2-7 Vernier calipers. Fine adjustment of the points is made by clamping the thumbscrews at the right and turning the knurled nut on the horizontal screw. *(Courtesy L.S. Starrett Company.)*

vernier is from the zero mark on the bar. Add the number of thousandths indicated by the line on the vernier that coincides with the line on the bar.

For example, in the Figure 2-8, the vernier has been moved to the right 1.000 plus 0.400 plus 0.025, which is equal to 1.425 inches, as

Figure 2-8 Detail of calipers with a vernier. This vernier permits readings to 0.001 inch. *(Courtesy L.S. Starrett Company.)*

shown on the bar. The eleventh line on the vernier coincides with a line on the bar, as indicated by the stars. Therefore, 0.011 inch is added to the reading on the bar, giving a total reading of 1.436 inches.

Bevel Protractors

A bevel protractor is the same as a bevel, but with a protractor added to it, which adapts it to all kinds of work in which angles are to be laid out. In general, there are two kinds of bevel protractors:

- Those without a vernier
- Those with a vernier

The bevel protractor without a vernier is suitable for angles that do not require a high degree of accuracy (Figure 2-9). The dial of the bevel protractor is accurately graduated from 0° to 90° to each extremity of an arc of 180°. It turns on a large central stud, which is hardened and ground, and can be clamped rigidly in any position after setting.

Figure 2-9 Bevel protractor. It turns on a large central stud, which is hardened and ground, and can be rigidly clamped in any position desired. The dial is accurately graduated in degrees over an arc of 180°, reading 0° to 90° from each extremity of the arc. *(Courtesy L.S. Starrett Company.)*

The *universal bevel protractor* with a vernier is graduated in degrees throughout the entire circle (Figure 2-10). The swivel turns on a large central stud, which is hardened and ground, and can be rigidly clamped by a thumbnut. The vernier increases materially the adaptability of the protractor for obtaining finer measurements. Readings to five minutes (5′), or 1/12 of a degree, can be obtained.

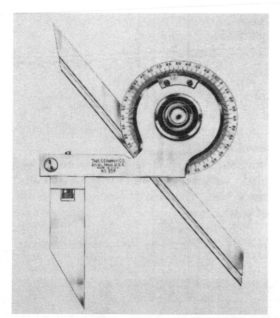

Figure 2-10 Universal bevel protractor with vernier. The vernier permits its use in obtaining fine measurements. *(Courtesy L.S. Starrett Company.)*

The dial of the protractor is graduated both to the right- and left-hand sides of 0° to 90°. The vernier scale is also graduated to the right and left of zero to 60 minutes (60′); each of the 12 vernier scales has graduations representing 5 minutes (5′). Any size angle can be measured because both the protractor dial and the vernier scale have graduations in opposite directions from zero (Figure 2-11).

Figure 2-11 Detail of a protractor vernier. The figures are close to the graduations to facilitate reading the vernier. *(Courtesy L.S. Starrett Company.)*

Since 12 graduations on the vernier scale occupy the same space as 23 degrees on the protractor dial, each vernier graduation is $\frac{1}{12}$ degree (or 5 minutes) shorter than 2 graduations on the protractor dial. Therefore, if the zero graduation on the vernier scale coincides with a graduation on the protractor dial, the reading is in exact degrees. However, if any other graduation on the vernier scale coincides with a protractor graduation, the number of vernier graduations must be multiplied by 5 minutes, and added to the number of degrees read between the zeros on the protractor dial and the vernier scale.

For example, in Figure 2-11, zero on the vernier scale is between "50" and "51" degrees on the protractor dial to the left of zero. Also, reading to the left, the fourth line on the vernier scale coincides with the "58" graduation on the protractor dial, as indicated by the stars. Therefore, 4 × 5 minutes (or 20 minutes) must be added to the number of degrees, giving a reading of 50 degrees and 20 minutes (50°20′).

Universal bevel protractors have several uses (Figure 2-12 and Figure 2-13). The blade with beveled ends enables measurement of

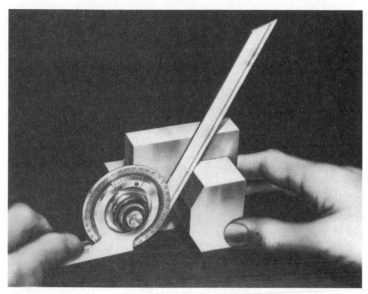

Figure 2-12   Uses of the universal bevel protractor. The blade with beveled ends permits measurements of angles from the vertex.
(Courtesy L.S. Starrett Company.)

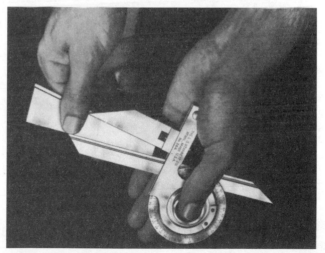

Figure 2-13   Universal bevel protractors may be used with parallels or knees for laying out work for inspection. The acute-angle attachment permits small angles. *(Courtesy L.S. Starrett Company.)*

an angle from the vertex. They may be used with parallels or knees for laying out work for inspection. An acute-angle attachment for quickly laying out small angles is available. One side of the tool is flat, which permits laying it flat on the work or paper.

Dial Indicators

A *dial indicator* (incorrectly called a *dial gage*) is an instrument for indicating size differences, rather than making measurements, as the dial indicator ordinarily is not used to indicate distance. A dial indicator can be used in combination with a micrometer to measure exact distances.

Variations in measurements are shown by the movement of a hand on the dial of the dial indicator. The dial is graduated in thousandths of an inch (that is, each division on the dial represents contact point movement of 0.001 inch).

The dial indicator (Figure 2-14) is useful in testing shafts for alignment, for checking cylinder bores for roundness and taper, and for testing bearing bores. The dial indicates the alignment (or roundness) of the piece tested to within 0.001 inch. A skilled workman can check alignment to within 0.00025 inch.

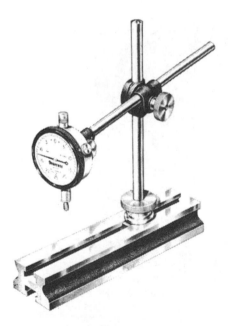

Figure 2-14 A dial test indicator with attachments.

(Courtesy L.S. Starrett Company.)

The dial indicator is extensively used in manufacturing and in service-and-repair work. Other uses are for straightening crankshafts, locating wrist-pin holes, determining the amount of shim to insert or remove, determining taper, checking play in bearings, reboring work, lining up magneto coil assembly, and so on (Figure 2-15).

Figure 2-15 A lathe operator using a dial indicator to center work.
(*Courtesy L.S. Starrett Company.*)

Another type of dial indicator is shown in Figure 2-16. This is a precision gage utilized on the precision dial comparator. It can also be used on a number of measuring gages.

Gages
A gage is often erroneously considered to be any measuring instrument. A gage is a *fixed* device that establishes a particular dimension, but it is not a measuring instrument.

However, some gages (a surface gage, for example) are adjustable and can be set to any desired dimension within their ranges. After it

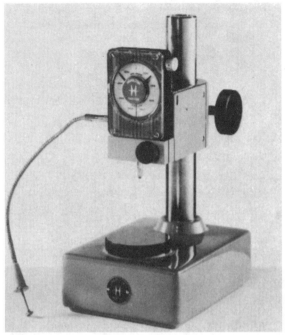

Figure 2-16 A precision dial comparator. This instrument is direct reading to 0.00005 inch. *(Courtesy Hamilton Watch Company.)*

is set to a particular dimension, a gage becomes a fixed device and is properly called a gage.

Surface Gage

A *surface gage* (Figure 2-17) is a machinist's instrument for testing planed surfaces. It has a heavy base, grooved through the bottom and end, adapting it for use on circular work as well as flat surfaces. The spindle may be set upright or at an angle, or turned to work under the base. It can be sensitively adjusted to any position.

Layout work often includes lines scribed at a given height from a face of the work, or a continuation of lines around several surfaces. It can be used to scribe lines at a given height on any number of pieces when duplicate parts are being made. Thus, the height of a standard bearing may be transferred to the faces of any number of castings from which duplicate bearings are to be made (Figure 2-18).

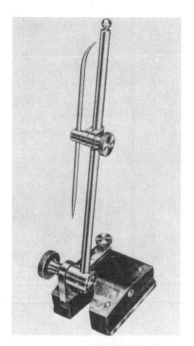

Figure 2-17   Universal surface gage.
(Courtesy L.S. Starrett Company.)

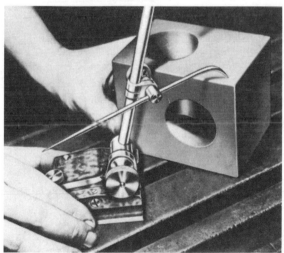

Figure 2-18   Machinist using a surface gage to level up a cast-iron block to determine the amount of work to be machined off.
(Courttesy L.S. Starrett Company.)

A clean surface plate and a combination square for obtaining the dimension are needed to set the surface gage properly (Figure 2-19). All instruments should be absolutely clean. The following three steps are necessary in setting a surface gage:

1. Adjust the standard to a convenient position.

2. Adjust the scriber to the approximate dimension.

3. Further adjust the scriber by turning the knurled adjusting screw on top of the base to the desired index line on the blade of the combination square.

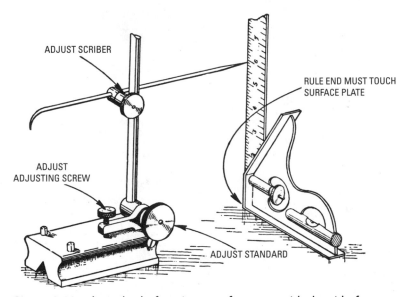

Figure 2-19 A method of setting a surface gage with the aid of a combination square. Make certain that the instruments and surface plates are absolutely clean. *(Courtesy L.S. Starrett Company.)*

Height Gage

The height gage is designed to measure or mark off vertical distances from a plane surface. The *vernier height gage* (Figure 2-20) is indispensable for layout, jig, and fixture making because of its fine adjustment, which permits extremely accurate measurements. The location of center distances of jigs, dies, and so on, can be accurately obtained by the use of toolmakers' buttons. A combination marker and extension may be used with the movable jaw for measuring and scribing lines on the work.

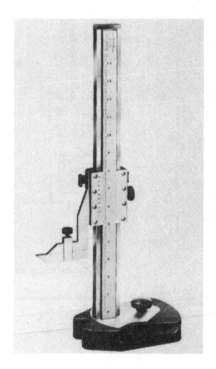

Figure 2-20   Vernier height gage.
(Courtesy L.S. Starrett Company.)

The end of the extension is beveled to a sharp edge for scribing lines (Figure 2-21).

The vernier height gage is graduated to read in thousandths of an inch, by means of a vernier scale on the sliding jaw. Graduations on one side are for outside measurements, and graduations on the other side are for inside measurements.

Depth Gage

The *vernier depth gage* is a similar precision instrument for measuring depths of slots, holes, and so on (Figure 2-22).

The *micrometer depth gage* is another accurate instrument used for vertical measurements. It is also essential for jig and fixture making (Figure 2-23). Depths of holes, slots, and so on, can be measured with micrometer accuracy.

Snap Gage

Snap limit gages are used for both *internal* and *external* dimensions (Figure 2-24). The distance between measuring surfaces is fixed and represents the size stamped on the gage. The gages are available in different sizes and can be used for measuring duplicate parts in

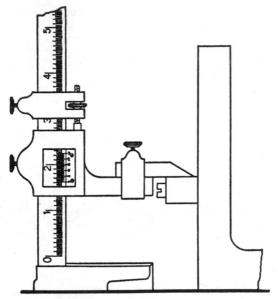

Figure 2-21 Vernier height gage. *(Courtesy L.S. Starrett Company.)*

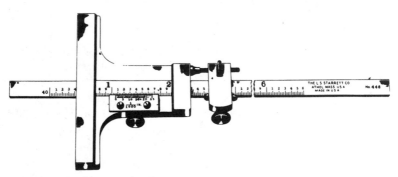

Figure 2-22 Vernier depth gage. *(Courtesy L.S. Starrett Company.)*

machine shop work. When these gages become worn, they can be closed in and reground, or lapped, to true size.

Plug Gage

Plug gages are also called *go* and *not go* gages (Figure 2-25). They have double ends, and have both a "go" end and a "not go" end (that is, when the work is at the correct size, one end of the gage will slip into it, but the other end will not). In some types of work, a

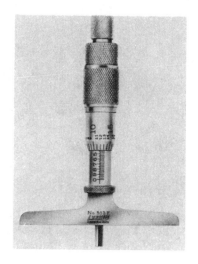

Figure 2-23   Micrometer depth gage. *(Courtesy Lufkin Rule Company.)*

given part may be required to be within certain size limits to be correct for a particular class of fit. A limit gage may be used for both the minimum and maximum sizes.

Ring Gage

Ring gages are standard cylindrical gages (male and female). The ring, or external, gage is a bored ring (Figure 2-26). It is also called a collar gage. These gages can be made to both "go" and "not go" dimensions.

Taper Gage

Taper gages are made of metal and have a graduated taper. There are two types of taper gages. One has a tapered thickness, and the other type has a tapered width.

The taper gage with tapered thickness (Figure 2-27) is used for bearing work and for gaging slots. The taper gage with the tapered width (Figure 2-28) is used as a gage for tubing.

Figure 2-24
Adjustable snap limit gage. The size is stamped on the gage. *(Courtesy Greenfield Tap & Die Company.)*

Figure 2-25   A plug gage. *(Courtesy Greenfield Tap & Die Company.)*

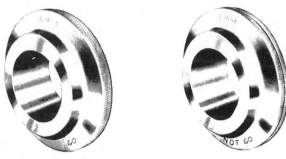

Figure 2-26 A ring gage. *(Courtesy Greenfield Tap & Die Company.)*

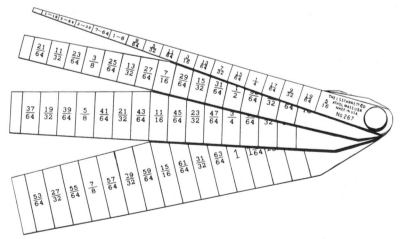

Figure 2-27 Taper gage with tapered thickness. Used for bearing work and for gaging slots. *(Courtesy L.S. Starrett Company.)*

Figure 2-28 Taper gage with tapered width. Used as a tubing gage.
(Courtesy L.S. Starrett Company.)

Center Gage

The center gage is a small tool that features the standard angle (60°) for lathe centers and for threading tools for the American National Standard screw threads. The center gage has a 60° point at one end and a 60° tool (Figure 2-29). It is used for testing angles of lathe centers and thread-cutting tools, and for setting tools at the correct angle relative to the work.

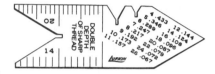

Figure 2-29 Center gage. The table on the gage is used for determining the size of tap drills for American National or U.S. Standard threads and gives the double depth of thread of tap and screw pitches.

(Courtesy Lufkin Rule Company.)

The large notch at the end of the center gage is used for testing lathe centers and threading-tool points. The small notches at the side of the gage are also used for testing tool points and for setting a threading tool at the correct angle to the work, by placing the opposite edge of the gage against the surface to be threaded. The tool is adjusted until the point fits into the notch in the gage.

The center gage may also be used for setting internal threading tools. The end of the gage is placed against the face of the work. The side notches are at right angles to the hole to be threaded, provided the work has been faced true.

Screw-Pitch Gage

The number of threads per inch, or pitch, of a screw or nut can be determined by the use of the screw-pitch gage (Figure 2-30). This device consists of a holder with a number of thin blades that have notches cut on them representing different numbers of threads per inch, the number being stamped on the blade. Some gages also have the double depth of thread (in decimals) stamped on the blade. This decimal number is equal to the depth of threads on the two sides of a tap. Thus, the workman can determine the size of tap drill to use in order to leave a full vee thread for a tap having the same pitch. To determine the size of drill needed, measure over the threads of the tap with a micrometer. From its size in thousandths of an inch, deduct the decimal number stamped on the pitch-gage leaf that agrees with the pitch of the tap to be used. The result is the correct size, in thousandths of an inch, of the drill needed for a full vee thread.

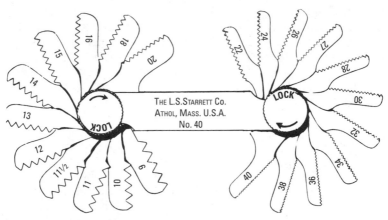

Figure 2-30 Screw-pitch gage. *(Courtesy L.S. Starrett Company.)*

Tapped threads need not be 100 percent full thread for commercial purposes. A tap drill that will give approximately a 75-percent thread is generally used. Sufficient stock in which to cut the threads must be left by the tap drill. A formula for finding approximate tap drill size is as follows:

Tap drill size = Major diameter of thread
$$- \frac{(0.75 \times 1.299)}{\text{No. threads per inch}}$$

A more practical and simpler formula is

$$\text{Tap drill size} = \text{Major diameter} - \frac{1}{\text{No. threads per inch}}$$

Tap drill sizes for full vee threads and for American National Standards threads can be calculated by the following:

- Full vee threads

$$d = D - \frac{1.733}{N}$$

- American National Standard Screw Threads

$$d = D - \frac{1.2999}{N}$$

In these formulas, D = major diameter of tap; d = minor diameter of tap; and N = number of threads per inch.

Tap and Drill Gage

The mechanic can use the *tap and drill gage* to select quickly the tap drill size for the tap to be used (Figure 2-31). The correct tap drill leaves enough stock to cut a full thread without breaking the tap; thus, uncertainty of result and much time can be saved.

Thickness or Feeler Gage

There are many varieties of *thickness*, or *feeler*, *gages* on the market. A thickness gage (Figure 2-32) consists of a number of thin steel

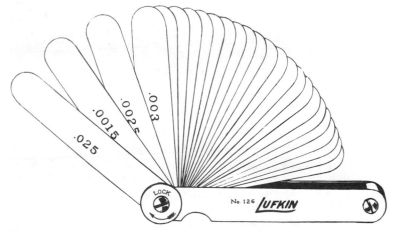

Figure 2-31   Tap and drill gage. *(Courtesy L. S. Starrett Company.)*

Figure 2-32   Thickness gage. The leaves of these gages include the most used sizes in automotive work. *(Courtesy Lufkin Rule Company.)*

leaves, which vary by thousandths of an inch. The leaves may be used singly or in groups. They enable the mechanic to form any desired thickness, within the limits of the tool. Standard thicknesses range from 0.0015 to 0.015, by thousandths of an inch.

Thickness (or feeler) gages are used extensively for setting ignition points, spark gaps, and valve tappets, and for checking ring clearances, piston clearances, etc. They are extremely valuable to the machinist and toolmaker for a variety of purposes.

Another form of thickness gage, the *feeler stock* (Figure 2-33), is also available. Feeler stock is accurate, high-grade, uniformly tempered, thickness gage stock. The size or thickness is marked on each piece in large, easily read figures. Feeler stock can be used in the same manner as the common thickness gage.

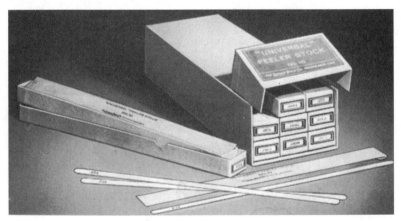

Figure 2-33   Feeler stock or thickness gages. *(Courtesy Lufkin Rule Company.)*

Wire Gage

The American Standard Wire Gage for nonferrous metals is shown in Figure 2-34. This gage is especially useful for electricians and others to gage sheets, plates, and wire made from the nonferrous metals (such as aluminum, brass, and copper). The decimal equivalents are stamped on the back of the gage.

Table 2-1 illustrates the various standards for wire gages. Stub's Iron Wire Gage is commonly known as the Birmingham gage and designates the Stub's soft-wire sizes. Stub's Steel Wire Gage is used to measure drawn steel wire or drill rod.

Figure 2-34   American Standard Wire Gage. It is generally standard for nonferrous metals such as copper, brass, and aluminum. These gages are useful for gaging sheets, plates, and wire. *(Courtesy L.S. Starrett Company.)*

U.S. Standard Gage for Sheet and Plate Iron and Steel

The U.S. Standard Gage for sheet and plate iron and steel is the recognized commercial standard in the United States for uncoated sheet, plate iron, and steel (Figure 2-35). It is based on weight in ounces per square foot. The decimal equivalents are on the back of the gage.

Effects of Temperature and Weight on Precision Tools

The larger micrometers are more sensitive to temperature changes than are the smaller sizes of micrometers. Where accurate measurements are to be taken with a large micrometer, the instrument should be tested with a pin gage for correct adjustment. In very large micrometers (24 inches to 36 inches), the micrometer, while being tested, should be held in the same position in which it is to be held when taking the measurement.

The weight of the frame of a micrometer may cause a variation in readings of the instrument. When using the larger micrometers in cold weather, it is also necessary to use a piece of waste or cloth between the hand and the frame. The heat transmitted from the hand may cause the frame to spring out of shape and cause a variation in readings.

Table 2-1 Standards for Wire Gages—Dimensions of Sizes in Decimal Parts of an Inch

No. of Wire	American or Brown & Sharpe for Nonferrous Metals	Birmingham or Stub's Iron Wire	American S. & W. Co.'s (Washburn & Moen) Std. Steel Wire	American S. & W. Co.'s Music Wire	Imperial Wire	Stub's Steel Wire	U.S. Std. Gage for Sheet & Plate Iron & Steel	No. of Wire
7-0s	0.651354		0.4900		0.500		0.500	7-0s
6-0s	0.580049		0.4615	0.004	0.464		0.46875	6-0s
5-0s	0.516549	0.500	0.4305	0.005	0.432		0.4375	5-0s
4-0s	0.460	0.454	0.3938	0.006	0.400		0.40625	4-0s
000	0.40964	0.425	0.3625	0.007	0.372		0.375	000
00	0.3648	0.380	0.3310	0.008	0.348		0.34375	00
0	0.32486	0.340	0.3065	0.009	0.324		0.3125	0
1	0.2893	0.300	0.2830	0.010	0.300	0.227	0.28125	1
2	0.25763	0.284	0.2625	0.011	0.276	0.219	0.265625	2
3	0.22942	0.259	0.2437	0.012	0.252	0.212	0.250	3
4	0.20431	0.238	0.2253	0.013	0.232	0.207	0.234375	4
5	0.18194	0.220	0.2070	0.014	0.212	0.204	0.21875	5
6	0.16202	0.203	0.1920	0.016	0.192	0.201	0.203125	6
7	0.14428	0.180	0.1770	0.018	0.176	0.199	0.1875	7
8	0.12849	0.165	0.1620	0.020	0.160	0.197	0.171875	8
9	0.11443	0.148	0.1483	0.022	0.144	0.194	0.15625	9

10	0.140625	0.191	0.128	0.024	0.1350	0.134	0.10189	10
11	0.125	0.188	0.116	0.026	0.1205	0.120	0.090742	11
12	0.109375	0.185	0.104	0.029	0.1055	0.109	0.080808	12
13	0.09375	0.182	0.092	0.031	0.0915	0.095	0.07196	13
14	0.078125	0.180	0.080	0.033	0.0800	0.083	0.064084	14
15	0.0703125	0.178	0.072	0.035	0.0720	0.072	0.057068	15
16	0.0625	0.175	0.064	0.037	0.0625	0.065	0.05082	16
17	0.05625	0.172	0.056	0.039	0.0540	0.058	0.045257	17
18	0.050	0.168	0.048	0.041	0.0475	0.049	0.040303	18
19	0.04375	0.164	0.040	0.043	0.0410	0.042	0.03589	19
20	0.0375	0.161	0.036	0.045	0.0348	0.035	0.031961	20
21	0.034375	0.157	0.032	0.047	0.0317	0.032	0.028462	21
22	0.03125	0.155	0.028	0.049	0.0286	0.028	0.025347	22
23	0.028125	0.153	0.024	0.051	0.0258	0.025	0.022571	23
24	0.025	0.151	0.022	0.055	0.0230	0.022	0.0201	24
25	0.021875	0.148	0.020	0.059	0.0204	0.020	0.0179	25
26	0.01875	0.146	0.018	0.063	0.0181	0.018	0.01594	26
27	0.0171875	0.143	0.0164	0.067	0.0173	0.016	0.014195	27
28	0.015625	0.139	0.0149	0.071	0.0162	0.014	0.012641	28
29	0.0140625	0.134	0.0136	0.075	0.0150	0.013	0.011257	29

(continued)

Table 2-1 (continued)

No. of Wire	American or Brown & Sharpe for Nonferrous Metals	Birmingham or Stub's Iron Wire	American S. & W. Co.'s (Washburn & Moen) Std. Steel Wire	American S. & W. Co.'s Music Wire	Imperial Wire	Stub's Steel Wire	U.S. Std. Gage for Sheet & Plate Iron & Steel	No. of Wire
30	0.010025	0.012	0.0140	0.080	0.0124	0.127	0.0125	30
31	0.008928	0.010	0.0132	0.085	0.0116	0.120	0.0109375	31
32	0.00795	0.009	0.0128	0.090	0.0108	0.115	0.01015625	32
33	0.00708	0.008	0.0118	0.095	0.0100	0.112	0.009375	33
34	0.006304	0.007	0.0104		0.0092	0.110	0.00859375	34
35	0.005614	0.005	0.0095		0.0084	0.108	0.0078125	35
36	0.005	0.004	0.0090		0.0076	0.106	0.00703125	36
37	0.004453		0.0085		0.0068	0.103	0.00664063	37
38	0.003965		0.0080		0.0060	0.101	0.00625	38
39	0.003531		0.0075		0.0052	0.099		39
40	0.003144		0.0070		0.0048	0.097		40

Figure 2-35   United States Standard Gage. This recognized commercial standard in the U.S. is used for uncoated sheet, plate iron, and steel. It is based on weights in ounces per square foot. Decimal equivalents of each gage number are stamped on the reverse side.

(Courtesy L.S. Starrett Company.)

Care should be taken not to press the two ends of the frame of large micrometers together. If this does occur and the effect is tested with a dial indicator of large-scale amplification, the importance of handling precision tools lightly may be demonstrated to a mechanic.

Electronic Caliper Depth Gages

Electronics make it easier to read the calipers inasmuch as interpretation is unnecessary with a digital readout. The electronic caliper depth gage is shown in Figure 2-36. It uses a liquid crystal display (LCD) to present the measurement in graduations of 0.0005 inch or 0.01 millimeter. It also has the option to convert from inches to millimeters. The zero-set button for zero can be set at any position in the range. The outside jaws measure up to 1.6 inch while the inside jaws are limited to 0.700 inch. Power is furnished by one silver oxide battery, which lasts for one year.

The electronic depth gage shown in Figure 2-37 can be used to measure diameters, depths, clearances, keyways, recesses, cross bores and groove widths. The digital readout is easy to read. All the

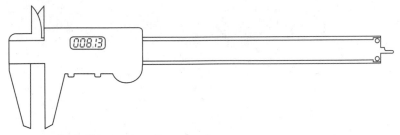

Figure 2-36 Electronic caliper depth gage.

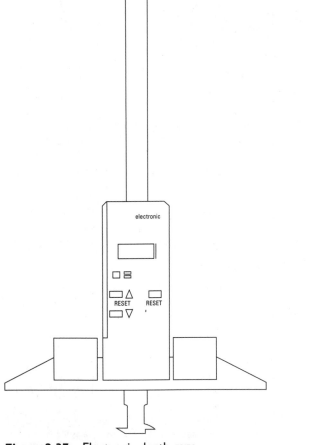

Figure 2-37 Electronic depth gage.

sliding and measuring surfaces have been lapped to within 0.001 mm
to give smooth movement without play, with constant contact pressure for accurate measurements. Repeatability is 0.01 millimeter or
0.0005 inch.

Another advantage of this model is its RS 232 serial output for
hookup to an interface for a computer and printer. The inch/mm
button instantly changes from inch to millimeter readings. You can
set the gage to zero in any position and read in either direction with
the correct sign displayed. There is no need to add or subtract values. Input the information with the appropriate sign and read off
the final result. The instrument has four different heads: groove-
measuring head, depth-measuring head, cross-bore–measuring
head, and universal measuring head. One of the main advantages of
this type gage is its design with industrial environments in mind. It
is not subject to wear and requires no maintenance. Figure 2-38
shows the groove head, depth head, cross-bore head and universal
head in use. Two models are available with 0- to 8-inch (200 millimeters) and 0- to 16-inch (400 millimeters) capabilities.

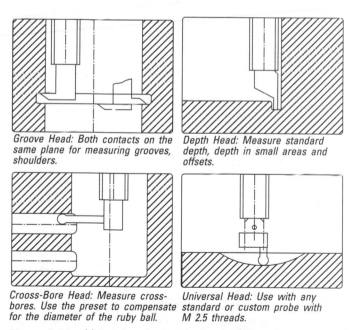

Groove Head: Both contacts on the
same plane for measuring grooves,
shoulders.

Depth Head: Measure standard
depth, depth in small areas and
offsets.

Crooss-Bore Head: Measure cross-
bores. Use the preset to compensate
for the diameter of the ruby ball.

Universal Head: Use with any
standard or custom probe with
M 2.5 threads.

Figure 2-38 Using the electronic depth gage with its groove head,
depth head, cross-bore head, and a universal head. *(Courtesy The Dyer Company.)*

Electronic Digital Micrometer

Some people have trouble reading a micrometer and handling the necessary math associated with the accuracy the device provides. The electronics field has expanded its reach to this area and provides the beginner or the person with difficulties reading a conventional micrometer an answer to the problem. The LCD screen is easily readable and the accuracy is the same as for the mechanical device.

The digital micrometer shown in Figure 2-39 is similar to the standard mechanical one, except that it reads out in decimals. There are a number of extra features with the LCD model. It has both inch and millimeter readings with the snap of a switch. It has a long-lasting, 3-volt (up to one year) battery and gives you the same accuracy as a standard mechanical micrometer with an automatic off after 30 minutes of non-use. The electronics unit can switch instantly from inches to millimeters, saving a lot of time and effort when making conversions. It also has a HOLD button for making sure the reading is available until properly eliminated by the operator. It has the ability to set minimum and maximum limits and the output data can be fed into a PC.

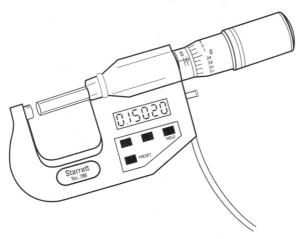

Figure 2-39 Electronic digital micrometer. *(Courtesy L.S. Starrett Company.)*

Summary

Many accurate measuring tools are used in the machine shop. Machinist measurements must be done with a much greater degree of precision than is required for many other lines of work. The measuring tools most commonly used are micrometers, verniers, calipers, protractors, indicators, and gages.

The word "micrometer" indicates a precision instrument for small measurements since "micro" means "extremely small" and "meter" is the unit of measurement in the metric system. The micrometer is a device that relies on the screw threads of its design to make extremely accurate measurements.

Micrometer calipers can be adjusted to compensate for wear. Some mechanics change a micrometer setting quickly by holding the sleeve and swinging the frame around several times. The vernier has ten divisions on the sleeve that occupy the same amount of space as nine divisions on the thimble. The difference between the width of one of the ten spaces on the vernier and one of the nine spaces on the thimble is one-tenth of a division on the thimble, or one-tenth of a thousandth of an inch, which is one ten-thousandth of an inch.

Vernier calipers have a graduated blade and an adjustable tongue. The bar of the calipers is graduated in fortieths (0.025 inch), with every fourth graduation being numbered to represent a tenth of an inch. The vernier scale is divided into 25 divisions, numbered 0, 5, 10, 20, and 25. The 25 divisions on the vernier scale occupy the same space as 24 divisions on the bar.

A bevel protractor is an instrument equivalent to a bevel, but with an added protractor. It adapts to all kinds of work in which angles are to be laid out. They are available with or without a vernier. The universal bevel protractor with a vernier is graduated in degrees throughout the entire circle. The dial of the protractor is graduated both to the right- and left-hand sides of 0° to 90°. Twelve graduations on the vernier scale occupy the same space as 23 degrees on the protractor dial. Each vernier graduation is $\frac{1}{12}°$ (or 5 minutes) shorter than 2 graduations on the protractor dial.

A dial indicator is an instrument for indicating size differences, rather than making measurements, as the dial indicator ordinarily is not used to indicate distance. It is useful in combination with a micrometer to measure exact distances.

A gage is often erroneously considered to be any measuring instrument. A gage is a fixed device that establishes a particular dimension, but is not a measuring device. It can be a surface gage used for testing planed surfaces or a height gage used to measure or mark off vertical distances from a plane surface. The depth gage is another accurate instrument used for vertical measurements. It is very important in making jigs and fixtures. The snap gage is used for both internal and external dimensions.

The plug gage is called a *go* and *not go* gage. It has double ends, and has a *go* end and a *not go* end (that is, when the work is at the correct size, one end of the gage will slip into it, but the other end will not). A ring gage is a standard cylindrical gage (male and female). The ring, or external gage, is a bored ring. It is also called a collar gage. These gages can be made to both *go* and *not go* dimensions.

Taper gages are made of metal and have a graduated taper. There are two types of taper gages. One has a tapered thickness, and the other type has a tapered width. The center gage is a small tool featuring the standard angle (60°) for both lathe centers and for threading tools for the American National Standard screw threads. The screw-pitch gage is used to determine the number of threads per inch, or pitch, of a screw or nut. It consists of a holder with a number of thin blades that have notches cut on them representing different numbers of threads per inch. The tap and drill gage is used to select a tap drill size quickly for the tap to be used. The correct tap drill leaves enough stock to cut a thread without breaking the tap.

The thickness or feeler gage comes in a variety of types. A thickness gage consists of a number of thin steel leaves, which vary by thousandths of an inch. The leaves may be used singly or in groups. They enable the machinists to form any desired thickness within the limits of the tool. The wire gage is called the American Standard Wire Gage for nonferrous metals. This gage is especially useful for electricians and others to gage sheets, plates, and wire made from the nonferrous metals (such as aluminum, brass, and copper). The decimal equivalents are stamped on the back of the gage. Stub's Steel Wire Gage is used to measure drawn steel wire or drill rod.

The larger micrometers are more sensitive to temperature changes than are the smaller sizes of micrometers. Where accurate measurements are to be taken with large micrometers, the instrument should be tested with a pin gage for correct adjustment.

Review Questions

1. Explain how to read a micrometer.
2. How is a vernier caliper used?
3. Describe the difference between bevel protractor and the universal bevel protractor.
4. How is the dial indicator used?
5. Name a few gages that are used by the machinist.

6. Precision measurements are generally written in decimals and read in thousandths and _____ of an inch.

7. What type of frame does a micrometer caliper have?

8. What does a bevel protractor do?

9. What is the purpose of a dial indicator?

10. Define what a gage does.

11. What is a snap gage?

12. For what is a plug gage used?

13. How are ring gages used?

14. What does the screw-pitch gage measure?

15. How is the wire gage used?

16. What is the other name for Stub's Iron Wire Gage?

17. How does the size of a No. 3 wire compare in an American or Brown & Sharpe gage and the U.S. Standard Gage for Sheet Metal & Plate Iron & Steel?

18. What effect does temperature and weight have on precision tools?

19. What does the term *micrometer* indicate?

20. What is a vernier?

Chapter 3

Materials

To facilitate efficient shop operations, every machinist and metal worker should have a general knowledge of the nature and properties of the materials with which he or she works.

Materials are generally classified as metallic or nonmetallic. Metallic materials are subdivided into a nonferrous category (such as copper, aluminum, or titanium) and a ferrous category (such as iron, steel, or various alloys). Nonmetallic materials include inorganic materials (such as ceramics, glass, and graphite) and organic materials (such as wood, rubber, and plastics).

Properties

A material is said to possess certain properties that define its character or behavior under various conditions.

Desirable Properties

Both *static strength* and *dynamic strength* are desirable properties in any material. Low cost is always desirable. Especially in the casting process, low cost may determine the material to be used, even though it may have some poor characteristics. For example, in cast metals, the following characteristics are desirable:

- Low melting temperature
- Good fluidity when melted
- A minimum of porosity
- Low reduction in volume during solidification (shrinkage)

Definition of Properties

Following are frequently used terms for expressing the properties of metals:

- *Brittle*—Breaks easily and suddenly with a comparatively smooth fracture; not tough or tenacious. This property usually increases with hardness. The hardest steel is the most brittle, and white cast iron is more brittle than gray iron. The brittleness of castings and malleable work is reduced by annealing and/or tempering.

- *Cold short*—The name given to a metal that cannot be worked under the hammer, or by rolling, or that cannot be bent when cold without cracking at the edges. Such a metal may be worked or bent at a high heat, but not at a temperature that is lower than dull red.

- *Cold shut*—In foundry work when, through cooling, the metal passing around the two sides of a mold does not properly unite at the point of meeting.

- *Ductile*—Easily drawn out, flexible, pliable. Material (such as iron) is "ductile" when it can be extended by pulling.

- *Elastic limit*—The greatest strain that a substance can endure and still completely spring back to the original shape when the strain is released.

- *Fusible*—Capable of being melted or liquified by the action of heat.

- *Hardness*—The ability to resist penetration or scratching.

- *Homogeneous*—Of the same kind or nature; as applied to boilerplates, homogenous means even-grained. In steel plates, there are no layers of fibers, and the metal is as strong one way as another.

- *Hot short*—More or less brittle when heated (such as hot-short iron).

- *Melting point of a solid*—The temperature at which solids become liquid or gaseous. All metals are liquid at temperatures more or less elevated, and they turn into gas (or vapor) at very high temperatures. Their melting points range from 39° below zero Fahrenheit (F) (the melting, or rather the freezing, point of mercury), up to more than 3000°.

- *Resilience*—The act or quality of elasticity; the property of springing back, or recoiling, upon removal of a pressure (as with a spring). Without special qualifications, the term is understood to mean the work given out by a spring or piece (stressed similarly to a spring) after being stressed to the extreme limit within which it may be stressed repeatedly without rupture or receiving *permanent set*.

- *Specific gravity*—The weight of a given substance relative to an equal bulk of some other substance, which is taken as a standard of comparison. Water is the standard for liquids and solids; air or hydrogen for gases. If a certain mass is weighed first in air and then in water, and the weight in air is divided by the loss of weight in water, the result is the specific gravity. For example, the weight of a 10-pound piece of cast iron suspended from a scale pan in a bucket of water is 8.6 pounds. Dividing 10 by the difference (10 minus 8.6, or 1.4), the specific gravity of cast iron is 7.14.

- *Strength*—Power to resist force; solidity or toughness; the quality of bodies by which they may endure the application of force without breaking or yielding.

- *Tensile strength*—The greatest longitudinal stress a substance can bear without tearing apart.

- *Toughness*—Having the ability to absorb energy without failure; capable of resisting great strain; able to sustain hard usage. Material (such as iron) is said to be "tough" when it can be bent first in one direction, and then in the other, without fracturing. The greater the angle it bends (coupled with the number of times it bends), the tougher it is.

Metals

A metal is any chemical element (such as iron, gold, or aluminum) that, when in a pure state and dissolved in an acid solution, carries a positive charge and seeks the negative pole in an electric cell. Metals are generally good conductors of heat and electricity, and are generally hard, heavy, and tenacious.

Ferrous Metals

The *ferrous metals* are those materials that contain iron. The machinist has long been concerned with the useful properties of iron.

Iron

Pure iron (ferrite) is a relatively soft element of crystalline structure. Pure iron solidifies at 2782°F, the temperature remaining at that point for a very short period of time, depending on the rate of cooling and the mass of the metal. Then, the temperature drops to 1648°F, where another pause occurs. On further cooling to 1416°F, the temperature again remains constant for a short time. No further pauses occur as it is cooled from 1416°F to atmospheric temperature.

Certain changes take place in pure iron as it cools. Pure iron can exist in four solid phases that have different physical characteristics. Following are forms of pure iron:

- *Alpha iron*—This iron is soft, magnetic, and incapable of dissolving carbon. Alpha iron occurs between atmospheric temperature and 1416°F.

- *Beta iron*—This phase of iron is feebly magnetic at the higher temperatures and nonmagnetic at the lower end of the range. It is intensely hard and brittle, and has almost no action on carbon. Beta iron occurs when it is heated from 1416°F to 1648°F.

- *Gamma iron*—In this phase, the iron readily takes up carbon, especially as temperature increases. If gamma iron is cooled quickly past the critical point, the passage of the hard gamma iron to the soft alpha iron is retarded. The iron is then in an unstable hardened condition, ready to pass into the soft alpha form of iron.

  The presence of many foreign substances (such as carbon, nickel, and manganese) seems to help gamma iron resist passing into alpha iron. Thus, hard gamma iron is more stable and permanent at lower temperatures.

  On the other hand, the presence of chromium, tungsten, aluminum, silicon, phosphorus, arsenic, and sulfur facilitates the passage of hard beta iron to the soft alpha form. As to hardness, gamma iron lies between that of alpha iron and beta iron. Gamma iron occurs between the temperatures of 1648°F and 2554°F.

- *Delta iron*—This form of iron has very little use. The liberation of heat at 2554°F indicates that, in changing from the delta to the gamma phase of iron, the internal structure of the metal has changed. The appearance of a critical point at 2554°F indicates that iron is in different states both above and below that point. Delta iron occurs between 2554°F and 2782°F.

Pig iron is a compound of iron with carbon, silicon, sulfur, phosphorus, and manganese. The carbon content of pig iron is from 2 to 4.5 percent. This occurs in two forms, partly in solution or combined and partly distributed throughout the mass in the form of graphite or uncombined carbon.

Cast iron generally cannot be formed and shaped by pressure, or rolled or drawn into shapes that are useful. It is remelted pig iron. The carbon content is over 2 percent, which indicates that it is not malleable at any temperature. Cast iron is widely used in industry for castings. Following are four types of cast iron:

- *Gray cast iron*—This is the kind most used in ordinary castings. This soft cast iron contains a high percentage of graphite, which renders it tough with low tensile strength. It breaks with a coarse-grained dark or grayish fracture. The color is caused by the presence of flattened flakes of graphite scattered throughout the material. If the flakes of graphite are large and numerous, the tensile strength is low. The size and amount of graphite flakes are determined by their formation during solidification. If solidification occurs rapidly, less carbon will separate as graphite. Therefore, there will be increased hardness because of increased amounts of combined carbon. Gray cast iron contains 2.5 to 3.5 percent carbon.

- *White cast iron*—Iron of somewhat lower carbon content (2.0 to 2.5 percent) is called white iron. This iron completely retains its combined carbon throughout the casting. Therefore, graphite is not formed, resulting in a casting of extreme hardness and brittleness. Where hardness is desired and brittleness can be tolerated, white cast iron can be used for machine parts.

  If some graphite formation has occurred, darker patches appear in the white iron. This indicates reduced hardness in those areas. Such an iron is called *mottled* iron.

- *Malleable cast iron*—Where it is desirable that complicated parts of a machine be ductile, malleable cast iron may be used rather than forged parts. Malleable cast iron castings may be bent or distorted (within the limits of the material) without breaking.

  Castings are made of hard, brittle, white cast iron and annealed (that is, converted into malleable iron). In the annealing process, the excess carbon is eliminated by heating in an extended heat treatment of about 1650°F for several hours. The carbon in the form of graphite is absorbed, converting the cast iron into a form of steel.

- *Wrought iron*—By definition, wrought iron is a low-carbon steel containing a considerable amount of slag. It differs from steel in the method of manufacture in that it is not entirely molten. It contains 1 to 2 percent of slag. White iron is used to produce wrought iron. The impurities are removed by the puddling process.

The presence of sulfur causes wrought iron to be brittle or "red short" when hot. The presence of phosphorus causes the iron to be "cold short" at ordinary temperatures. Wrought iron softens and welds at 1600°F. It can be forged at still lower temperatures.

Steel

This is a general term that describes a series of alloys in which iron is the base metal and carbon is the most important added element. The simple steels are alloys of iron and carbon, containing no other elements, and ranging from 0 to 2.0 percent carbon. The "pure" steels (iron and carbon alloy) have never been made in quantity.

The commercial *plain carbon steels* contain manganese as a third alloying element, and small quantities of silicon, phosphorus, sulfur, and traces of other elements. The term plain carbon steel is used for steels containing from a few hundredths of a percent carbon to 1.4 percent carbon. The properties depend on both the carbon content and the heat treatment.

Low-carbon steels are ordinarily used either in the "rolled" condition or in the annealed condition, while high-carbon steels are used where extreme hardness is desired. Increasing the content of steel up to a certain percentage increases its strength. Beyond that point the strength decreases. For example, mild steel containing 0.1 percent carbon has a tensile strength of about 50,000 pounds per square inch. A carbon content of 1.2 percent increases the tenacity to nearly 140,000 pounds per square inch, which is probably the limit for carbon steel. A 2.0 percent carbon content gives it a tensile strength of about 90,000 pounds per square inch. A further gradual increase in carbon content causes the material to rapidly acquire the characteristics of cast iron.

Plain carbon steels seldom contain more than 1.4 percent carbon. A carbon content of 2.0 percent is the theoretical upper limit. Various elements other than carbon are added to give steel its desired properties. The effects of adding these elements are as follows:

- Phosphorus enhances the hardness of steel and makes it better able to resist abrasion. Steel high in phosphorus is weak against shocks and vibratory stresses. Thus, phosphorus is considered a harmful impurity in steel boilerplates.

- Sulfur interferes with the shaping and forging of steel because it increases the brittleness of steel while hot, making it "red short." Sulfur content of steel should not exceed 0.02 to 0.05 percent.

- Manganese increases the strength, hardness, and soundness of steel. If a considerable proportion of manganese is present, steel acquires a peculiar brittleness and hardness that makes it difficult to cut. It has a neutralizing effect on sulfur.

- Nickel increases both the strength and toughness of steel.
- Aluminum improves the soundness of ingots and castings.
- Vanadium renders steel nonfatigable. It gives great ductility, high tensile strength, and high elastic limit, making the steel highly resistant to shocks.

  Vanadium steels contain 0.16 to 0.25 percent vanadium. These steels are specially adapted for springs, car axles, gears subjected to severe service, and for all parts that must withstand constant vibration and varying stresses.

  Vanadium steels containing chromium are used for many automobile parts (such as springs, car axles, drive shafts, and gears). Most chrome-vanadium steels contain 0.20 to 0.60 percent carbon. Many heat-treated forgings are made from these steels.

- Molybdenum is sometimes specified for high-speed steel. Molybdenum steel is suitable for large crankshafts and propeller shafts, large guns, rifle barrels, and boilerplates.

High-speed steel is so named because it can remove metal faster when used for cutting tools, as in the lathe. The high-speed steel retains its hardness at higher temperatures. Such tools may operate satisfactorily at speeds that cause the edges to reach a red heat.

These steels contain 12 to 20 percent tungsten, 2 to 3 percent chromium, usually 1 to 2 percent vanadium, and sometimes cobalt. The carbon content is within rather narrow limits (usually 0.65 to 0.75 percent). The most used steel is referred to as 18-4-1 steel, which means that its content is 18 percent tungsten, 4 percent chromium, and 1 percent vanadium. Another favorite is 14-4-2 steel. Steels containing 18 percent tungsten are best for most purposes, but steels lower in tungsten content are somewhat cheaper.

Stainless steel resists oxidation and corrosion when correctly heat treated and finished. It is not absolutely corrosion-resistant. Stainless cutlery and surgical and dental instruments contain 12 to 14 percent chromium.

Cast steel used in steel castings is stronger than cast iron. Steel castings made of stainless steel resist oxidation at temperatures up to 1800°F or higher, depending on the chromium content.

Cast steel does not pour as sharply as iron. The shrinkage for steel castings is greater than that for iron castings because of the high temperature of pouring the steel.

Nonferrous Metals

Nonferrous metals are metals other than iron. In addition to the metals already mentioned, the important nonferrous metals are copper, zinc, tin, antimony, lead, and aluminum.

Copper

Copper is one of the most useful metals in its pure form and in its various alloys (such as brass or bronze). Copper has a brownish-red color and is both ductile and malleable. It is very tenacious and one of the best conductors of heat and electricity.

Pure copper melts at 1980°F; commercial copper melts at 1940°F. The strength of copper decreases rapidly with a rise of temperature above 400°F; its strength is reduced to about half between 800° and 900°F. The heat conductivity of copper is greater than that of all other metals, except silver. It is also similar to silver in electrical conductivity.

Zinc

In the form of ingots, zinc is called spelter. Zinc is brittle at ordinary temperatures; it is ductile and malleable between 212° and 300°F, and it again becomes brittle at 410°F.

Zinc is used for lining cisterns and for coating iron water pipes or sheet iron for making cisterns. When the water with which it comes in contact is soft and contains a slight acid, the metal is gradually corroded or eaten away. Zinc tarnishes when exposed to moist air and is corroded when in contact with soot and moisture. Zinc is used for eaves, gutters, and so on. When rolled into thin sheets, zinc is useful on roofs because of its lightness and ease of handling.

Tin

The melting point of tin is 450°F. It is ductile and easily drawn into wire at 212°F (the boiling point of water). Tin has a low tenacity, but it is very malleable and can be rolled into very thin sheets. Tin is used as a protective coating for iron and copper. It is also used for lining lead pipes designed for conveying drinking water because of its high resistance to tarnishing when exposed to air and moisture.

Antimony

Antimony is a hard brittle metal that resembles tin. It combines readily with other metals, forming alloys that are extensively used commercially.

Lead

Other metals have largely replaced lead in the plumbing industry. Lead is the heaviest of the common metals; it melts at 621°F. Lead is soft enough to be cut with a knife. It is malleable and ductile, but compared with other metals, it is not a good conductor of heat or electricity. Lead has a low tensile strength. It is extensively used to alloy with other metals for bearings and solders.

Bismuth
This is a remarkable metal because of two properties: Its specific gravity decreases under pressure, and it expands on cooling. The melting point is about 520°F. Bismuth is frequently used with antimony in type metals because it fills the molds completely on solidification.

Aluminum
This metal is the lightest of the common metals. Aluminum occurs in nature in the form of hydrates and silicates, but it is commercially prepared by the aid of electricity from cryolite and bauxite.

The metal is not corroded by atmospheric influences or fresh water, and also resists nitric acid. However, it is decomposed by alkalies in seawater and by dilute sulfuric acid. Aluminum is malleable, ductile, and a good conductor of heat and electricity. Thermal expansion of aluminum is slightly more than twice that of steel and cast iron.

Refractory Metals
Tantalum, tungsten, and molybdenum are commonly called *refractory metals* because their melting points are above 3632°F (2000°C). Because of their high melting points and their reactivity at extremely high temperatures, these metals are produced by powder metallurgy rather than by smelting. Hydraulic presses compact the metal powders to form bars of suitable size and shape for further processing. The bars are fragile and resemble chalk when they are removed from the press. However, they are made into strong metal by the sintering process, which consists of heating the bars in furnaces from which the air is excluded. This causes the particles of powder to fuse together without actually melting, and starts a regular metallic crystal growth.

Improvement and development of forming, fabricating, and welding techniques have increased the use of these metals. Examples of these achievements are drawn seamless tubing and the adaptation of the inert gas arc welding process to make welding of tantalum and molybdenum relatively easy.

Tungsten and Molybdenum
The most extensively used metals of the ten refractory metals are tungsten and molybdenum. Tungsten has the highest melting point

of all the metals. Most of the uses of these two metals are due to their high melting points and their ability to retain strength and stiffness at high temperatures.

Electronic tubes use tungsten for filaments, heaters, anodes, and seals through glass; molybdenum is used for grids, anodes, and support members. Tungsten and molybdenum are used for electrical contacts in automotive ignition, vibrators, telegraph relays, and other devices in which the contact parts are in practically continuous service. The pure metals may be used in contacts, but they may be used in combination with silver or copper to form metals having high arc-resisting qualities.

Tungsten is the principal ingredient of Fansteel 77 Metal, a heavy material that approaches tungsten in density but is easily machinable. Fansteel 77 Metal is used for rotors, flywheels, balance weights, and other rotational control parts where maximum weight or inertia, accompanied by high strength, is required for installations with limited space.

Both tungsten and molybdenum are used as heating elements in electric furnaces where working temperatures of 1600° to 2000°C are required. This is above the range of nickel-chrome alloys. Siliconized coating has been developed to permit the use of molybdenum heating elements up to 3000°F (1650°C) .

Tungsten electrodes are used to maintain the arc in inert gas of atomic hydrogen welding. Consumption of tungsten electrodes is very low because of the high melting point and low vapor pressure of the metal when used in helium, argon, or hydrogen atmospheres.

Tungsten and molybdenum heating elements are used in vacuum equipment for deposition of thin metallic or nonmetallic coatings by vaporization. Products coated in this manner are mirrors, television tubes, headlight reflectors, photographic lenses, and many similar items.

Tungsten and molybdenum are available in square and rectangular bars, sheet and plate, rods, wire, and metal powder. Molybdenum seamless tubing is available in a wide range of diameters and wall thicknesses. Tungsten carbide powder is also available.

Tantalum

Tantalum is an element best known for its almost complete resistance to corrosion and chemical attack. Very few acids have even the slightest effect on tantalum.

Tantalum has the ability to immobilize residual gases in electronic tubes at high temperatures. Other desirable properties for this purpose are its high melting point, low vapor pressure, thermal expansion, and ease of fabrication.

A third important property of tantalum is its ability to form highly stable anodic films. This action, combined with immunity to the corrosive action of electrolytes, is the basis of tantalum rectifiers, arresters, and electrolytic capacitors.

Columbium is a sister metal and occurs in the same ores with tantalum. It has properties generally similar to tantalum, although in a lesser degree.

Most of the metals with high melting points tend to be brittle and difficult to form or fabricate. Tantalum and columbium are exceptions. They are as malleable and ductile as mild steel, and all drawing, rolling, and forming operations are performed with the cold metal.

Because of its easy workability and its immunity to corrosion, tantalum has been widely used for surgical implants in the human body. Sutures are made of braided or monofilament wire. Severed nerves are repaired with fine wire and foil. Tantalum plate is used to repair skull injuries, and woven tantalum gauze is used in hernia operations.

Tantalum capacitors have long been used in telephone service, and the recent trend toward miniature components, along with the advent of television and other electronic applications, has greatly increased the use for tantalum capacitors. These capacitors are made from either sheet or foil.

Tantalum and columbium are available as square or round bars, rods, sheet and plate, foil, wire, powder, and as carbides. Tantalum tubing is available in a wide range of diameters as seamless, butt-welded, or seam-welded.

Tantalum equipment is recommended for operations involving chlorine or its compounds, including hydrochloric acid. It will not react with bromine, iodine, or other compounds. It has extensive use with sulfuric or nitric acids, hydrogen peroxide, and a large number of other acids, organic or inorganic compounds, and salts under conditions in which any other metal would be corroded quickly or would contaminate the purity of the processed material.

Nonferrous Alloys

A *nonferrous alloy* is a mixture of two or more metals containing no iron. The result of such a mixture is generally a mechanical mixture, but it may be combined chemically. With respect to properties, the mixture may be regarded as forming a new metal. The number of possible alloys is unlimited. Some of the more important alloys are considered here.

Brass

There are many varieties of brass. It is a yellow alloy composed of copper and zinc in various proportions. Small percentages of tin, lead, and other metals are included in some varieties of brass. In general, the composition of brass is determined by its desired color. The percentage of zinc in the various varieties of brass is as follows: red (5 percent), bronze color (10 percent), light orange (15 percent), greenish yellow (20 percent), yellow (30 percent), and yellowish white (60 percent).

Brass may be classified as (1) high brass or (2) low brass, meaning that the alloy has a high or low percentage of copper. The so-called brasses contain 30 to 40 percent zinc, being suitable for cold rolling. The low brasses contain 37 to 45 percent zinc and are suitable for hot rolling.

The commercial brasses are given various degrees of hardness by cold rolling, being designated as quarter-hard, half-hard, and full-hard. Tensile strength is variable, according to composition and treatment.

Bronze

Bronze is a nonferrous alloy of copper and tin. Many special bronzes have other ingredients. The greater the proportion of tin above 5 percent, the more brittle the alloy becomes.

Bronze is used instead of brass in some instances because of its better appearance and greater strength. A 1 to 6 percentage of tin is specified for bronze that is to be rolled cold and drawn into wire. If it is to be worked at red heat, 6 to 15 percent tin is specified. The percentages of tin specified for the following uses are machine parts, 9 to 20; bell bronze, 20 to 30; and art bronze, 3 to 10. There are a number of special bronzes, such as phosphor, manganese, gun, and tobin bronze, the properties of which differ, adapting them to special uses.

Aluminum

Although there is no limit to the number of alloys of aluminum that could be produced, commercial manufacturing considerations require that the number of alloys be as small as possible to provide the necessary combinations of properties to meet the needs of industry.

The elements commonly used in the production of casting alloys of aluminum are copper, silicon, magnesium, nickel, iron, zinc, and manganese. The strength of aluminum may be increased by addition of proper amounts of some of these elements. For example, alloys containing magnesium in suitable proportions, as the hardener, are even more resistant to corrosion than the aluminum-silicon alloys.

Aluminum alloy castings are poured both in sand molds and in permanent metal molds. In addition, certain alloys are cast in pressure die-casting machines, which also use metal molds or dies. Permanent molds, or dies, are practical only where a large number of identical castings are required. The minimum number that will justify the production of a metal mold, or die, varies greatly with the nature of the casting.

Babbitt Metal

Discovered in 1839 by a goldsmith from Boston named Issac Babbitt, Babitt metal is an alloy of tin, antimony, and copper. The United States granted Babbitt $20,000 for the right to use his formula in government work, and Massachusetts Charitable Mechanics Association awarded him a gold medal in 1841. Babbitt's formula is a good one. Unfortunately, competition and high-priced materials have encouraged adulteration, and the genuine formula is not always followed unless the alloy is subject to chemical analysis.

Other Nonferrous Alloys

Some of these alloys are used for special purposes in industry.

Monel metal is an alloy of copper and nickel and a small percentage of iron. Its melting point is 2480°F, and it may be forged at 165° to 1100°F. An important use for monel metal is in ship propellers.

Muntz metal is an alloy containing 60 percent copper and 40 percent tin. It can be used for purposes in which a hard sheet brass is desirable.

Tobin bronze is an alloy containing 58 to 60 percent copper, about 40 percent zinc, and a small percentage of iron, tin, and lead. Its tensile strength is about 60,000 pounds per square inch.

Delta metal is similar in composition and properties to Tobin bronze.

White metal is a term applied to various alloys containing mainly zinc and tin, or zinc, tin, and lead. It is used for bearings.

Tantung is a trade name for a series of alloys that have great hardness, strength, and toughness, and resistance to wear, heat, impact, corrosion, and erosion, even at extremely high temperatures. These alloys are composed chiefly of cobalt, chromium, and tungsten, with either tantalum or columbium carbide and other components added. A carbide of either tantalum or columbium imparts a low coefficient of friction, a self-lubricating action which minimizes wear.

One of these alloys, *Tantung G*, is widely used in tipped lathe tools and milling cutters, as well as in solid bits. It is available in rods and bars that can be used as tool bits or converted into punches, rollers, drills, and other special tools or wear-resisting parts.

Nonmetals

A *plastic* is a nonmetallic material that can be readily molded into intricate shapes. There are hundreds of plastic products on the market, and new plastic products are being developed continually.

Bakelite was one of the early products, and is used for many purposes. It is a phenol resinoid and has high mechanical resistance.

Formica is a laminated plastic product having many uses. It can be purchased in almost any size and shape desired—tubing, bars, rods, sheets, and so on.

Many other plastic products are on the market (such as acetate, vinyl, nylon, polyethylene, *Teflon*, and many others). All of these products have properties that adapt them to either a specific item or a variety of items.

Tests of Materials

A material's purpose determines the kinds of properties that the material must possess. Steel used in construction of bridges, buildings, and certain types of machines should possess strength, toughness, and elasticity as desirable properties. In some kinds of tool manufacture, a high degree of hardness, in addition to strength and toughness, may be required. The mechanical properties of metals may be changed by using alloy elements and by heat treatment. Specific tests have been devised to evaluate the properties of metals so that materials can be selected wisely.

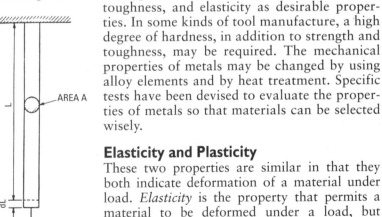

Elasticity and Plasticity

These two properties are similar in that they both indicate deformation of a material under load. *Elasticity* is the property that permits a material to be deformed under a load, but causes it to return to its original shape when the load is removed (Figure 3-1).

Plasticity is similar to elasticity in that the material is deformed under load, but the material retains some of the deformation after the load has been removed (Figure 3-2). This indicates that the

Figure 3-1 Application of stress to a metal rod.

elastic limit of the material has been exceeded.

Stress and Strain

Stress is the load per unit area (pounds per square inch). If the weight (W) has caused the metal rod to stretch a small amount, indicated by dl, the rod will return to its original length (L) when the weight is removed if the metal has elasticity. However, if the bar is longer than the original length after the weight has been removed, the metal has exhibited plasticity (Figure 3-1).

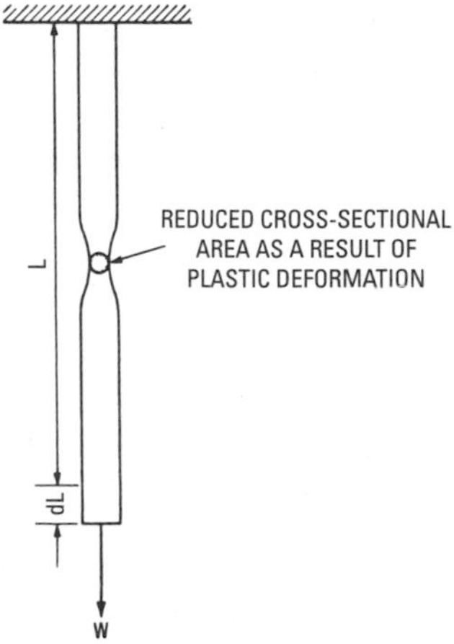

If W is increased further and causes a reduced cross-sectional area in the rod, the rod may break at this point. Thus, strain is the deformation per unit of length (measured in inches per inch) that results from a given stress (Figure 3-2).

REDUCED CROSS-SECTIONAL AREA AS A RESULT OF PLASTIC DEFORMATION

Figure 3-2 A metal rod stressed beyond the elastic limit.

This relationship between stress and deformation is known as the *elastic limit*. There is an elastic limit for all materials—a point beyond which complete recovery after stressing is not possible. Thus, the material is permanently deformed.

Tensile Strength

Testing machines for tensile strength have been designed so that gradually increasing loads may be applied to the material, together with apparatuses for measuring the corresponding deformation. Some of these machines automatically plot the relationship between stress and strain.

Ductility

The percent elongation is a measure of *ductility*. This property is the ability of a metal to permit plastic elongation.

Toughness

Toughness is the ability of a metal to assume deformation without rupture. Medium-carbon steel has a higher degree of toughness than high-carbon steel. However, the maximum strength of high-carbon steel is greater than that of medium-carbon steel.

Hardness

Hardness is a property that predicts the behavior of the tested material. There are three main types of hardness: penetration hardness, wear hardness, and rebound hardness.

Commercial hardness testers are widely used for testing metals. They are widely used for inspection and correlation with other properties (such as plasticity, toughness, and tensile strength), and in the specifications of materials for a definite purpose or use.

Portable Hardness Tester

The new Starrett portable hardness tester is an advanced, integrated unit with a very compact size, high accuracy, a wide measuring range, and simplicity of operation (Figure 3-3). It is ideal for testing the hardness of all metals in many areas of industry. It combines a universal impact device and a data processor in a single unit. With the ability to "plug and play" optional impact devices, this unit can effectively test any combination of material, size, or shape. It automatically computes at Vickers, Brinell, Rockwell, and Shore hardness values. Statistical mean value is automatically provided. It is battery-powered with a built-in D impact probe with an internal tungsten carbide test tip. Parts can be checked at any angle with highly accurate results, and then the data are placed in archives. It has an error modification function and automatic shut off.

Brinell Hardness Test

The Brinell hardness tester (Figure 3-4) has been widely used. A 10-mm hardened steel ball is pressed into the material under a hydraulic load. Depending on the material being tested, a measured load (3000-kilogram load for iron and steel) is applied for a short time interval (not less than 15 seconds). The diameter of the indentation of the ball is then measured in two directions. The *Brinell hardness number* is a ratio of the applied load and the surface area of the indentation of the ball in the material being tested.

The Brinell hardness test may be considered a destructive test for some conditions: the softer the metal, the larger the indentation of the ball, and the lower the Brinell hardness number.

Rockwell Hardness Test

The operation of the Rockwell hardness tester is similar to the Brinell tester in that an applied load presses a penetrator into the metal being tested (Figure 3-5). In this tester, a hardened steel ball is used for some tests and a Brale (or diamond-tipped cone) is used on materials too hard to be tested with a ball. Also, a minor load is first applied and then a major load. The minor load produces an initial indentation. The

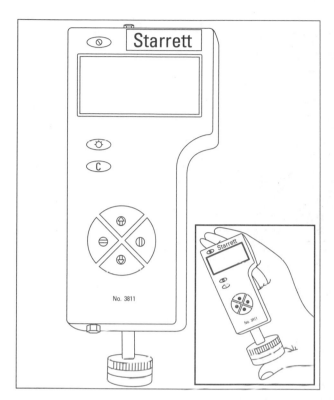

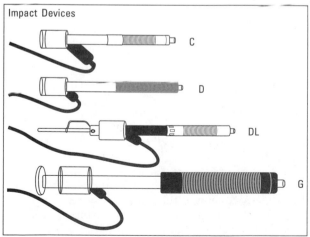

Figure 3-3 Portable electronic hardness tester. *(Courtesy L.S. Starrett Company.)*

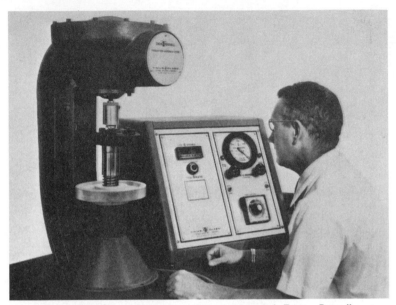

Figure 3-4 Air-o-Brinell metal hardness tester with Digito-Brinell system for digital readout of Brinell values. *(Courtesy Tinius Olsen Testing Machine Company.)*

dial is then set at zero, and the major load is applied for a time interval. Hardness numbers can be read directly from the indicating dial.

The Rockwell test is classified as nondestructive because the indentations are so small that it can be used for finished articles. It can be performed rapidly, and the accuracy is good. In the higher hardness range, it is considered more accurate than the Brinell test.

Shore Scleroscope

The Shore scleroscope determines hardness by dropping a small diamond-pointed hammer on the surface of the material being tested and measuring the height of the rebound (Figure 3-6). The hardness number is expressed in terms of the rebound distance. For a standard test, the specifications are given for the hammer. The height of rebound is given on a scale. The corresponding number on the scale is the *scleroscope hardness number*. The softer the material, the greater the deformation caused by the striking hammer, and the less energy available for rebound.

The machine is portable and provides a nondestructive method for hardness testing. However, the precautions that must be observed may be a disadvantage for certain conditions.

Figure 3-5 Rockwell hardness tester. *(Courtesy Wilson Mechanical Instrument Div. of American Chain & Cable Company.)*

File Hardness

File hardness is the oldest of the hardness tests. When a sharp file is drawn slowly and firmly across the surface, the material is considered "file hard" if the file does not "bite" into the surface. If the file does bite into the surface, the material is considered softer than file hard. If the file cuts quickly and easily into the surface, the material is soft.

The disadvantages of the file test lie in the fact that there are differences in the files used and in the operators, and in the fact that the hardness cannot be recorded as numerical data. The advantages, of course, are that the test is cheap, rapid, and nondestructive. A skilled inspector may be able to use the test to discard

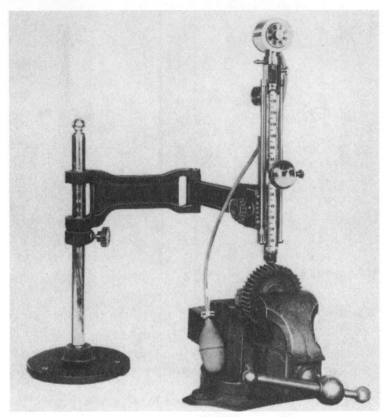

Figure 3-6 Shore scleroscope. *(Courtesy The Shore Instrument & Mfg. Company.)*

unsatisfactory pieces without the use of more expensive and sensitive equipment.

Brittleness
Brittleness is considered the opposite of toughness. A brittle material can undergo little or no plastic deformation. As the hardness of a material increases, the brittleness increases.

Relationship between Mechanical Properties and Hardness
The mechanical properties (such as elasticity, toughness, plasticity, ductility, and tensile strength) may be indicated by hardness tests of the metal. Figure 3-7 shows one type of hardness tester. Thus, the prediction of these properties may be of as much value to the engineer as the actual hardness value of the metal.

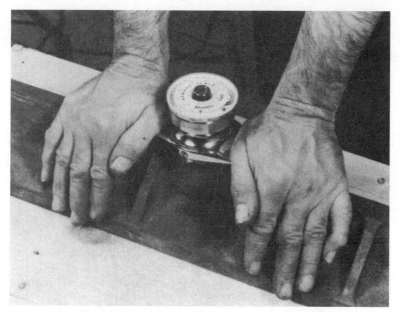

Figure 3-7 Portable hardness tester. *(Courtesy Newage Industries, Inc.)*

The degree of hardness of a given metal affects the mechanical properties as follows:

- Toughness can be expected to increase as hardness decreases. In general, toughness decreases as plasticity and ductility decrease.
- Plasticity of a metal increases as hardness decreases. When hardness of a metal becomes great enough, the metal will rupture before plastic deformation takes place.
- Ductility of a metal decreases as hardness increases. A decrease in ductility may make fabrication of a metal more difficult.
- Tensile strength increases as hardness increases.

Effects of Temperature
All the mechanical properties mentioned are affected by temperature. As temperature increases, tensile strength and hardness decrease. As temperature increases, plasticity and deformation also increase. The effects of temperature on the mechanical properties of

metals provide the real basis for study of the entire subject of heat treatment of metals.

Summary

Materials most commonly used in machine shop work are iron, steel, alloys, and plastic. Materials possess certain properties that define their character or behavior under various conditions. Both static strength and dynamic strength are desirable properties in any material. Cost is also a factor and, in many cases, may determine the material to be used.

A metal such as iron, gold, or aluminum is a chemical element. Metals are generally good conductors of heat and electricity, and are generally hard, heavy, and tenacious. Desirable properties are low melting temperature, good fluidity when melted, a minimum of porosity, and low reduction in volume during solidification (shrinking).

Terms frequently used to express the properties of metals are brittle, cold short, cold shut, ductile, elastic limit, fusible, hardness, homogeneous, hot short, melting point of a solid, resilience, specific gravity, strength, tensile strength, and toughness.

Ferrous metals are those that contain iron. There are many forms of pure iron known by Greek names such as alpha, beta, gamma, and delta. Pig iron is a compound of iron with carbon, silicon, sulfur, phosphorus, and manganese. Gray cast iron is the kind most used in ordinary castings. It contains a high percentage of graphite and contains from 2.5 to 3.5 percent carbon-flattened graphite, which gives it a gray color. There is also white cast iron, malleable cast iron, and wrought iron. Wrought iron has a low carbon content and melts at 1600°F.

Steel is a general term that is used to describe a series of alloys in which iron is the base metal and carbon is the most important added element. Simple steels are alloys of iron and carbon, with no other elements, that range from 0 to 2.0 percent carbon. The "pure" steels have never been made in quantity. Phosphorus, sulfur, manganese, nickel, aluminum, and vanadium are added to the mix to make steels for various applications.

High-speed steel is so named because it can remove metal faster when used for cutting tools, as in the lathe. Molybdenum is sometimes specified for high-speed steel. Nonferrous metals are metals other than iron. Important nonferrous metals are copper, zinc, tin, antimony, lead, and aluminum. Nonferrous alloys are a mixture of two or more metals, containing no iron. Brass is one of these nonferrous mixtures. Brass has a different name for different percentages of zinc contained (e.g., red has 5 percent zinc; bronze color has 10 percent zinc; and yellowish white has 60 percent zinc). Brass may be classified as high brass or low brass, depending on whether the alloy contains a high or low percentage of copper. Many special bronzes have other ingredients than copper and tin or copper and zinc.

There are other nonferrous metals such as Monel metal, Muntz metal, Tobin metal, delta metal, white metal, Tantung, and Tantung G. The last one is widely used to make special wear-resisting tools (such as lathe tools and milling cutters). Plastic is a nonmetallic material that can be readily molded into intricate shapes. Formica is a laminated plastic, while Bakelite is one of the early plastics. Today there are acetates, vinyls, nylons, polyethylenes, and Teflon, plus many others. Elasticity and plasticity are two properties that are similar in that they both indicate deformation of a material while under load.

Elasticity is the property that permits a material to be deformed under a load, but causes it to return to its original shape when the load is removed. Plasticity is similar to elasticity in that the material is deformed under load, but the material retains some of the deformation after the load has been removed. Stress and strain are also important when working with metal. Stress is the load per unit area or pounds per square inch. Testing machines have been designed to measure the relationship of stress and strain on metal.

There are a number of properties important to the working of metals. They are the metal's ductility, toughness, hardness, and brittleness. All of these properties are affected by temperature.

Review Questions

1. Materials are classified as _____ or _____.
2. List four desirable properties of materials.
3. Define the following terms:
   a. Brittle
   b. Cold short
   c. Cold shut
   d. Ductile
   e. Elastic limit
   f. Fusible
   g. Hardness
   h. Homogeneous
   i. Toughness

4. A _____ is a chemical element, such as iron, gold, or aluminum.

5. Ferrous materials are those materials that contain _____.

6. What are the four solid phases that pure iron can take?

7. What is pig iron?

8. What is cast iron?

9. What is the desirable characteristic of malleable cast iron?

10. What is wrought iron?

11. What is the meaning of the word "steel"?

12. What is in plain carbon steels to make them commercially important?

13. Plain carbon steels seldom contain more than ____ percent of carbon.

14. What effect does each of the following have when added to steel?

   a. Phosphorus

   b. Manganese

   c. Sulfur

   d. Aluminum

   e. Nickel

   f. Vanadium

15. How does high-speed steel get its name?

16. What is cast steel?

17. What is stainless steel?

18. At what temperature does copper melt?

19. What is the melting point of tin?

20. What are the two properties of bismuth?

21. Where is aluminum obtained?

22. Name three refractory metals.

23. What is columbium?

24. What is meant by the term *nonferrous*?

25. List the percentage of zinc in the varieties of brass listed below.

   a. Red

   b. Bronze color

    c. Light orange

    d. Greenish yellow

    e. Yellow

26. What is bronze?

27. What is Babbitt?

28. What is Monel metal?

29. Identify the properties of the following metals:

    a. Muntz metal

    b. Tobin bronze

    c. White metal

    d. Tantung

30. Identify the following nonmetals:

    a. Formica

    b. Teflon

    c. Bakelite

31. What is elasticity? Plasticity?

32. Why is tensile strength important?

33. What is ductility?

34. Describe toughness and hardness.

35. What is the oldest of the hardness tests?

36. What is the property of metal that is the opposite of hardness?

37. All mechanical properties of metals are affected by _____.

Chapter 4

Abrasives

Grinding, as we know it today, is not an old art. Most of the development in grinding has taken place since 1890. By definition, an *abrasive* is a substance (such as sandpaper or emery) that is used for grinding, polishing, and so on.

Structure of Abrasives

For centuries the only abrasives used were prepared from sandstone. With the development of harder metals and alloys, particularly steel, harder and more efficient abrasives were needed.

Natural Abrasives

Emery (Figure 4-1) and corundum have been known for a long time as hard, natural minerals. They are naturally occurring forms of aluminum oxide, corundum having the larger crystals and containing fewer impurities. These readily available natural abrasives were used in manufactured grinding wheels until after the turn of the century, when they were superseded by more efficient, man-made abrasives. An almost negligible quantity of natural abrasives is in use in grinding wheels today (Figure 4-2).

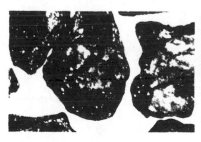

(A) Turkish emery has a higher aluminum oxide content.

(B) American emery.

Figure 4-1   Emery is a natural mixture of aluminum oxide and magnetite (Fe^3O^4). It is often preferred to synthetic abrasive for coated abrasive use. Emery is ideal for nonskid surfaces in conjunction with concrete or epoxy resins. It is also used for barrel finishing, pressure blasting, and general polishing operations.

(Courtesy American Abrasive Company.)

(A) Silicon carbide abrasive.

(C) Regular aluminum oxide abrasive.

(B) Bauxite, as mined.

(D) A modified form of aluminum oxide abrasive. Pure white in color, and contains about 98.6 percent pure aluminum oxide.

Figure 4-2 Examples of bauxite and fused abrasive as broken from the pig. *(Courtesy Norton Company.)*

The principal sources of emery are Turkey, Greece, and Asia Minor. Corundum first came from India, and later deposits were found in South Africa, Canada, and the southern part of the United States. Shipments from the various localities, and even from the same locality, differed so widely in quality that it caused a variation in the quality of the wheels produced.

Manufactured Abrasives

As grinding became more refined, variation in quality presented a serious problem, which led to experiments in producing manufactured abrasives whose quality could be controlled.

Glue, shellac, and silicate of soda were first used as bonding materials to hold the abrasive grains together in the grinding wheel. Progress in the development of bonding materials was not made until the ceramic clays and firing kilns of the potter were adapted to grinding wheel manufacture.

Composition of Abrasives

Formerly, all grinding wheels were made of emery. The cutting element in emery is crystalline aluminum oxide. New abrasives have replaced emery in grinding wheels, although it is still used for some forms of grinding.

Silicon Carbide

The commercial use of silicon carbide abrasives was developed by the Carborundum Company. In 1891, Edward G. Acheson, an electrical engineer who lived in Monongahela, Pennsylvania produced a few ounces of small-sized bright crystals from a mixture of clay and powdered coke that he had heated in a small, crude electric furnace. He found that these crystals would scratch glass like a diamond. Chemical analysis showed the crystals to be silicon carbide (Figure 4-2A). These crystals were sold for polishing precious gems, at a cost approaching that of the gems. The Carborundum Company developed the use of silicon carbide abrasive for grinding from this discovery.

Aluminum Oxide

Around the same time that silicon carbide abrasive was first produced, Charles B. Jacobs, chief engineer for Ampere Electro-Chemical Company at Ampere, New Jersey, set out to make a synthetic corundum. *Bauxite* (Figure 4-2B) was fused electrically into a hard material (crystalline aluminum oxide) similar to emery and corundum, but it had an advantage in that it could be produced in uniform grade and was of much higher purity (93 to 94 percent aluminum oxide). Its use in grinding wheels and as a polishing abrasive was developed by the Norton Company. Today, approximately 75 percent of all grinding wheels are made with an aluminum oxide abrasive of one type or another, making it the most widely used abrasive for grinding wheels (Figure 4-2C and Figure 4-2D).

Diamonds

Actual synthesis of the diamond was achieved in 1955 by the General Electric Company, although the basic principles of the diamond's formation, heat and pressure, had been known for many years. Diamonds are the hardest materials found in nature, and attempts had been made for many years to reproduce them in laboratories.

Natural diamonds were developed for use in grinding wheels in the early 1930s to grind tungsten carbide, a material so hard that it resists the abrasive action of ordinary grinding wheels. The development of man-made diamonds for grinding wheels since 1959 has ensured that industry will have a reliable supply and freedom from price fluctuations of natural diamonds. Diamond wheels are widely used today for grinding carbide, ceramics, glass, stone, and even some tool steels (Figure 4-3).

Figure 4-3   Grinding a carbide cutter with a diamond grinding wheel.
(Courtesy Norton Company.)

Use of Abrasives in Grinding Wheels

Grinding is the process of disintegrating a material and reducing it into small particles of dust by crushing or attrition. Many of the newer abrasives have been developed for special grinding purposes.

Silicon Carbide Abrasives

Silicon carbide is made from pure silica sand and carbon, in the form of finely ground coke. These materials react when subjected to the high temperatures of an electric furnace to form silicon carbide crystals, which are widely used today in a large variety of abrasive and refractory products (Figure 4-4). Following are some of the brand names of silicon carbide abrasives:

- Crystolon
- Carborundum

- Carbolon
- Carbonite

Figure 4-4   Black silicon carbide. Used for grinding and finishing nonferrous and nonmetallic materials, lapping and polishing, and in several coated abrasive products. *(Courtesy American Abrasive Company.)*

Aluminum Oxide Abrasives

Bauxite is the source of aluminum oxide, Al_2O_3. Bauxite is aluminum combined with water and varying quantities of impurities (Figure 4-5).

Figure 4-5   Regular aluminum oxide. This is a tough abrasive, well suited for general-purpose grinding, nonskid surfaces, barrel finishing, lapping, pressure blasting, and polishing. *(Courtesy American Abrasive Company.)*

Arc-type electric furnaces are used for the manufacture of aluminous abrasives. The finished product is a large "pig" of crystalline aluminum oxide. Following are some brand names of the aluminum oxide abrasives:

- Alundum
- Aloxite
- Lionite

- Borolon
- Exolon

Diamond Abrasive

As the abrasive in bonded grinding wheels, crushed and sized diamonds came into use in the early 1930s for cutting tools, wear-resistant products, molded boron carbide, quartz, crystal, glass, porcelain, ceramics, marble, and granite. Diamond abrasive particles in grinding wheels enable the wheels to retain their shape, although the particles will dull eventually.

Both natural and man-made diamonds have fields of application in which they excel. However, the advent of man-made diamond abrasive has given users of diamond abrasives a wider selection of diamond grinding wheels adapted to specific uses. Diamond grinding wheels are made in the following three different bond types:

- *Resinoid*—The resinoid bonded diamond wheel is characterized by a very fast and cool cutting action.
- *Metal*—The metal bonded diamond grinding wheel has unusual durability and high resistance to grooving.
- *Vitrified*—The vitrified bonded diamond grinding wheel has a cutting action comparable to that of the resinoid diamond wheels and a durability approaching that of the metal-bonded diamond wheels.

The conventional manufactured abrasives (silicon carbide and aluminum oxide) are essential for the needs of many industries, not only for grinding and polishing operations, but also for use in other final abrasive products. They are the basic ingredients in grinding wheels, oilstones, and pulpstones, and abrasive-coated paper and cloth. They are also used in nonslip floors, stair tile, porous plates, refractories, and refractory laboratory ware.

Summary

Very few natural abrasives are used in industry because of the harder metals and alloys produced. Manufactured abrasives are used in grinding because uniformity and composition can be controlled. Aluminum oxide is the most widely used abrasive for the manufacture of grinding wheels.

Synthetic diamonds for use in grinding wheels were developed in 1959. Diamond wheels are widely used today for grinding carbide, ceramics, glass, stone, and even some steel tools.

Silicon carbide is another product used for grinding. It is made from pure silica sand and carbon. These materials react when subjected to the high temperatures of an electric furnace to form the silicon carbide crystals widely used in various abrasives.

Natural abrasives were used long before the more structured materials were discovered. Emery and corundum were used for years because they are naturally occurring forms of aluminum oxide. With the development of harder materials, the need for a better form of grinding stone was highlighted. Grinding is one of the important steps in producing a finished product made of iron, steel, or other hard material.

Review Questions

1. Why are manufactured abrasives used in industry?
2. What is the most popular abrasive material used in industry?
3. Why are synthetic diamonds used in grinding or cutting tools?
4. What is an abrasive?
5. Where do you use abrasives in metalworking?
6. What is meant by bonding in a grinding wheel?
7. How is black silicon carbide used in the finishing of metal?
8. What is the source of aluminum oxide?
9. Why are diamonds used in the finishing of metals?
10. What does *vitrified* mean?
11. Define *abrasive*.
12. What is emery used for?
13. What is corundum?
14. What is magnetite?
15. List four examples of bauxite and fused abrasive.
16. How was silicon carbide developed?
17. Where are diamonds used for grinding purposes?
18. Why are diamonds used for grinding purposes?
19. Define *grinding*.
20. List four brand names for silicon carbide abrasives.
21. How are aluminous abrasives manufactured?
22. List four brand names for aluminum oxide abrasives.
23. What are the three different bond types of diamond grinding wheels?
24. What is the difference between emery and sandpaper?
25. Who was Edward G. Acheson?

Chapter 5

Grinding

By definition, *grinding* is the process of disintegrating a material and reducing it into small particles of dust by crushing or attrition.

Manufacture of Grinding Wheels

Grinding wheels are manufactured either from sandstone, which occurs in nature, or from man-made abrasives. Sandstone was the chief abrasive material for many years.

Natural Grindstones

The natural grindstones are cut from sandstone, the most common of which is the Berea sandstone of Ohio. Natural grindstones are inexpensive and have been used largely in the glass, cutler, and woodworking edge-tool industries. Manufactured abrasive wheels, however, have largely replaced natural grindstones in these plants because these grinding wheels can be operated at higher speeds, and the grain size, hardness, and structure can be controlled.

Manufactured Abrasive Grinding Wheels

Most of these grinding wheels are made from aluminum oxide and silicon carbide. They are used in industries requiring modern high-speed grinding equipment.

Preparation of the Abrasive

The aluminum oxide and the silicon carbide abrasive wheels are prepared in a similar manner. The ore from the electric furnace is rough-crushed to lumps about 6 inches in diameter. It is then shipped to the abrasive mill, where the lumps are reduced to about ¾-inch diameter (or finer) by powerful crushers. Finally, the pieces are reduced in size to grains suitable for use in grinding wheels, coated abrasive products, and so on, by passing them through a series of steel crushing rolls.

Any iron impurities that are present are removed by conveying the abrasive grains through magnetic separators. All fine abrasive dust and foreign particles are removed by washing thoroughly with steam and hot water. This operation is important because clean abrasive grains mix more uniformly with the bond. The abrasive is then dried in continuous rotary driers. The abrasive grains are accurately sized, or graded, by passing them over a series of vibrating

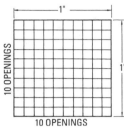

Figure 5-1 Abrasive grain sizing.

screens. Standard grain size numerals range from 10 to 600, which refer to the "mesh" of the screen through which a particular grain size will pass. A grain that will pass through a screen having 10 mesh openings per linear inch is called a 10-grit size. This 10-grit screen therefore will have 100 openings per square inch (Figure 5-1).

The abrasive grain, as it comes from the mill, is inspected for capillarity, uniformity of size, strength, and weight per unit volume. This represents the finished abrasive product of the mill, and it is either transported directly to the various grinding wheel manufacturing departments or stored in huge tanks for later use.

Abrasive grains are uniformly distributed throughout the bond in the wheel. The structure of the grinding wheel refers to this relative spacing as dense, medium, or open, depending on the percentage of abrasive or pores (Figure 5-2). Wheels of medium structure are best for hard, dense materials; open structures are best for heavy cuts and for soft, ductile materials that are easily penetrated and require good chip clearance.

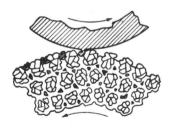

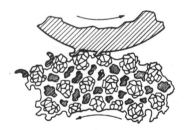

(A) Proper spacing provides effective chip clearance.

(B) Too-close spacing causes the wheel to load easily, and it will not cut effectively.

Figure 5-2 Correct spacing of the abrasive grains within the grinding wheel is important for fast free-cutting action.

(Courtesy Cincinnati Milacron Company.)

Shapes of Grinding Wheels

Grinding wheel manufacturers have standardized 9 shapes and 12 faces for grinding wheels (Figure 5-3 and Figure 5-4). These wheels are made in a wide range of sizes to do most grinding jobs.

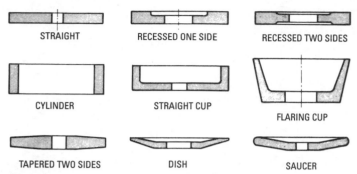

Figure 5-3 Nine standard shapes for grinding wheels. These shapes will perform most jobs. *(Courtesy Cincinnati Milacron Company.)*

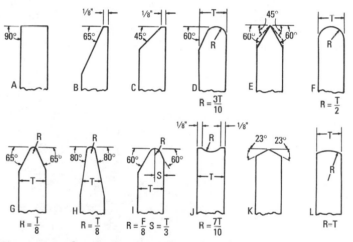

Figure 5-4 Standardized grinding-wheel faces. These faces can be modified by dressing to suit the needs of the user.

(Courtesy Cincinnati Milacron Company.)

A great variety of "mounted pieces" and mounted wheels are made for precision grinding of small holes and for offhand grinding, as on dies (Figure 5-5). These mounted wheels and "mounted points" vary in size from $\frac{1}{16}$ to $1\frac{1}{2}$ inches in diameter.

Method of Mounting Grinding Wheels
Grinding wheels should be checked for balance before mounting. Out-of-balance wheels set up excessive vibration, causing chatter

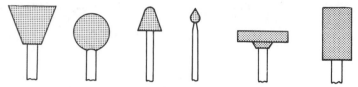

Figure 5-5 Mounted points are tiny grinding wheels permanently mounted on small-diameter shanks. They may be as small as $1/16$ inch in diameter. *(Courtesy Cincinnati Milacron Company.)*

marks on the ground surface and excessive wear on the bearings and spindle.

Wheels with small holes are held by a single nut on the end of the spindle. This nut should be tightened firmly, but not excessively, to avoid setting up excessive strains in the wheel. Large-hole wheels are held in place by flange screws around the sleeve. These should be drawn tight with the fingers. Then, the diametrically opposite screws should be tightened with a wrench (preferably a torque wrench) until all screws have been tightened uniformly, but not excessively. These screws should be checked occasionally for looseness. Grinding wheels should be mounted correctly for safe operation (Figure 5-6).

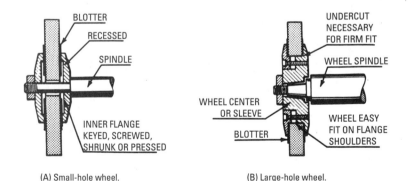

(A) Small-hole wheel.   (B) Large-hole wheel.

Figure 5-6 Correct method of mounting a grinding wheel. Never omit the blotting-paper washers. *(Courtesy Cincinnati Milacron Company.)*

Truing and Dressing the Grinding Wheel
A wheel should have enough of the cutting face removed, in preparation for grinding, to have it running true with its own spindle. When a wheel becomes loaded or glazed, it is then "dressed" to restore its original sharp and clean cutting face.

Three chief types of wheel dressers are in use on precision-grinding machines. These are the diamond tool, abrasive wheel, and mechanical dresser. The diamond tool is most commonly used on all kinds of precision grinders (Figure 5-7). All types of dressers must be held in firm mounts on the machine to true or dress a wheel accurately. This may be either an integral part of the machine or a special attachment for the machine. In many automatic machines, wheel dressing is part of the automatic cycle.

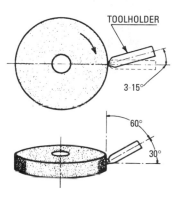

Figure 5-7 Diamond truing tools must be canted, as shown, to prevent chatter and gouging of the wheel, and to maintain sharpness of the diamond abrasive. *(Courtesy Cincinnati Milacron Company.)*

Bond Types in Grinding Wheels

The bond in a grinding wheel holds the grains together and supports them while they cut. The greater the amount of bond with respect to the abrasive, the heavier the coating of bond around the abrasive grains and the "harder" the wheel. The abrasive itself is extremely hard in all grinding wheels. The terms *hard* and *soft* actually refer to the strength of the bonding of the wheel (Figure 5-8). The general types of grinding wheel bonds are

- Vitrified
- Organic (resinoid, rubber, shellac)
- Silicate

Vitrified-Bond Grinding Wheels

About two-thirds of all the grinding wheels manufactured are made with a vitrified bond composed of clays and feldspars, and selected for their fusibility (Figure 5-9). During the "burning" process of manufacture, a temperature of 1270°C is reached, which is

Figure 5-8   Abrasive grain and bond are mixed in power mixing machines to ensure a homogeneous blending. *(Courtesy Norton Company.)*

sufficiently high to fuse the clays into a molten glass condition. Upon cooling, a span or post of this glass connects each abrasive grain to its neighboring grains to make a rigid, strong grinding wheel (Figure 5-10).

Organic-Bonded Grinding Wheels

Organic-bonded grinding wheels are made with an organic bond such as resinoid, rubber, or shellac. Resinoid wheels are used primarily in high-speed, rough-grinding operations and make up by far the largest percentage of organic-bonding grinding wheels. Straight wheels are used widely on bench and pedestal grinders for offhand grinding of castings. Cup wheels and cones are used for cleaning castings and for weld grinding with portable grinders. Reinforced cutoff wheels are used on both cutting-off machines and portable grinders to remove gates and risers from casting, for cutting bar stock, and for cutting concrete blocks and other masonry materials.

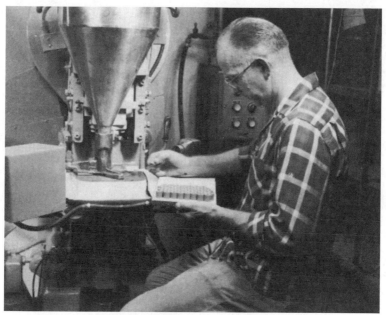

Figure 5-9 Small grinding wheels are molded in automatic presses. The operator merely removes the pressed wheel between strokes. An electric-eye safety device stops the machine if his hand is under the plunger on the pressing stroke. *(Courtesy Norton Company.)*

Rubber-bonded grinding wheels are used as regulating wheels in centerless grinders to control the travel of the workpiece through the machine (Figure 5-11). Rubber cutoff wheels are used where there must be a minimum burr left by the wheel. They should always be used with a good flow of coolant, while resinoid-bonded cutoff wheels can be used dry. Resilient rubber wheels are used in the hollow grinding of cutlery, as well as on work where finish is important.

Shellac-bonded wheels find their greatest uses in grinding the rolls used for making steel, paper, plastics, and so on (Figure 5-12). Shellac-bonded cutoff wheels are used to some extent in tool rooms for cutting off the ends of broken taps and drills. Burning of the heat-treated tool steel must be avoided and wheel wear is of minor importance.

Figure 5-10 Grinding wheels emerging from an electrically heated tunnel kiln where the approximately 2500°F temperature has fused the clay bond into a strong glass. *(Courtesy Norton Company.)*

Figure 5-11 Rubber-bonded wheels are used for the regulating wheels on centerless grinding machines; also for thin cutoff wheels. *(Courtesy Norton Company.)*

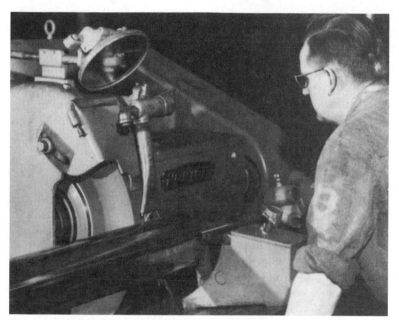

Figure 5-12   Shellac-bonded wheels are widely used for cutlery grinding and high-finish grinding. *(Courtesy Norton Company.)*

Silicate-Bonded Grinding Wheels

These wheels derive their name from the bond, which is principally silicate of soda. Silicate-bonded wheels are considered relatively "mild acting" and are still used to some extent in the form of large wheels for grinding edge tools in place of the old-fashioned sandstone wheels.

Grinding Wheel Markings

A standard system for marking grinding wheels was adopted in 1944 by the various grinding wheel manufacturers throughout the country. Each marking consists of six parts, arranged in the following sequence:

- Abrasive type
- Grain size
- Grade
- Structure
- Bond type
- Bond modification symbol

Abrasive Type

Manufactured abrasives fall into two distinct groups. Letter symbols are used to identify these two groups as follows:

- A—Aluminum oxide
- C—Silicon carbide

To designate a particular type of either silicon carbide or aluminum oxide abrasive, the manufacturer may use its own symbol or brand designation as a prefix. Examples include Norton Company's 32A, 57A, 37C, or 39C.

Grain Size

Grain size in wheels is indicated by numbers ranging from 10 (coarse) to 600 (fine). If necessary, the manufacturer may use an additional symbol to the regular grain size. This is illustrated by the examples 461 (46 grit, No. 1 combination) and 364 (36 grit, No. 4 combination).

Grade

The grade of a grinding wheel is indicated by a letter of the alphabet ranging from A to Z (soft to hard) in all bonds or manufacturing processes. Grades A to H are soft; grades I to P are medium; and grades Q to Z are hard.

Structure

Structure (or grain) spacing in a grinding wheel is indicated by a number, generally from 1 to 12. The progressively higher numbers indicate a progressively "more open" or wider grain spacing (Figure 5-13).

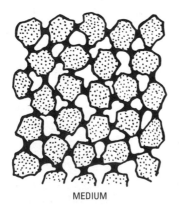 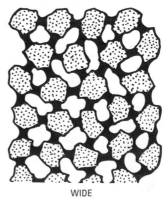

MEDIUM   WIDE

Figure 5-13 Grinding-wheel structures showing medium (left) and wide (right) grain spacings. *(Courtesy Norton Company.)*

For example, a 3 structure indicates a dense or close grain structure; an 8 structure indicates a wide grain spacing. The letter "P" at the end indicates a "P" type product having extra large pores or voids. Examples include 32A30–E12VBEP and 39C601–H8VKP.

Bond or Process

The type of bond is designated by the following letters:

- V—Vitrified
- B—Resinoid (synthetic resins such as Bakelite)
- R—Rubber
- E—Shellac
- S—Silicate

Bond Modification Symbols

Bond modification symbols may describe a particular type of bond that distinguishes it as a variation from a basic bond. The position may be either omitted or shown according to the characteristics of the specified wheel. Examples include VG (Norton "G" type of vitrified bond) and B11 (Norton "11" type of resinoid bond).

Factors Affecting Grinding Wheel Selection

Selection of a grinding wheel for a given operation depends on a number of important factors. Some grinding wheels are designed for special purposes (Figure 5-14).

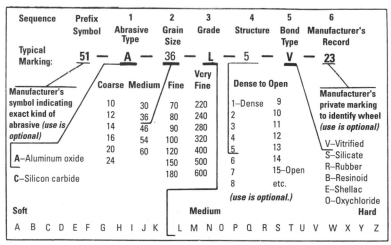

Figure 5-14 Standard grinding-wheel marking system.

(Courtesy Cincinnati Milacron Company.)

Hardness of Material to Be Ground
The kind of abrasive to be selected is determined by the nature and properties of the material to be ground.

Abrasive
Aluminum oxide is best suited for grinding steel and steel alloys. Silicon carbide grinding wheels are more efficient for grinding cast iron, nonferrous metals, and nonmetallic materials.

Grit Size
Fine grit is best for hard, brittle, and difficult-to-penetrate materials. Coarse grit is best for soft, ductile, easily penetrated materials.

Grade
A relatively soft-grade grinding wheel is required for very hard, dense materials. Hard materials resist the penetration of the abrasive grains and cause them to dull quickly; a soft-grade wheel enables worn, dull grains to break away and expose newer, sharper cutting grains. Harder-grade grinding wheels should be used for soft, easily cut materials.

Amount of Stock to Be Removed and Finish Required
Grit size and bond are important in the selection of a wheel, depending on the amount of stock to be removed and the finish required.

Grit Size
A coarse grit is best for rough grinding or rapid stock removal. A fine grit size is best where close tolerances and a high finish are desired.

Bond
The vitrified bond is best suited for fast cutting and a commercial finish. A resinoid, rubber, or shellac bond is usually best for obtaining a high finish.

Operation (Wet or Dry)
As a rule, wet grinding permits the use of grinding wheels at least one grade higher than for dry grinding without danger of burning the work from heat of friction. Water speeds up the work and reduces dust.

Wheel Speed
The grinding wheel speed (Figure 5-15) is very important in the selection of a wheel and bond as follows:

- For speeds less than 6500 surface feet per minute (sfpm), use vitrified-bonded wheels.
- For speeds above 6500 sfpm, use organic-bonded wheels (resinoid, rubber, or shellac).

Figure 5-15 As a safety precaution, almost all grinding wheels, 6 inches and larger, are speed tested at 50 percent higher than their recommended operating speeds. *(Courtesy Norton Company.)*

Area of Grinding Contact

Grit size and grade are influenced by the area of contact between the wheel and the work. As a general rule, the smaller the area of contact, the harder the grinding wheel should be. Therefore, a coarse-grit wheel can be used for a large area of contact, while a fine-grit wheel should be used where the area of contact is small.

Severity of Grinding Operation

This factor affects the choice of the abrasive as follows:

- Use a tough, abrasive, "regular" aluminum oxide for grinding steel and steel alloys under severe conditions. The use of a tougher abrasive is limited generally to grinding stainless steel billets and slabs with heavy-duty grinders.
- Use a relatively mild abrasive for light grinding on hard steels.
- Use an intermediate abrasive for a grinding job of average severity.

Summary

Grinding refers to the shaping or smoothing of the surfaces of various materials and the shaping of cutting tools. The two basic abrasive materials used in manufacturing grinding wheels are silicone carbide and aluminum oxide. Silicon carbide is harder than aluminum oxide but is more brittle. Abrasive grains are uniformly distributed throughout the bond in a grinding wheel. The structure of the grinding wheel refers to this relative spacing as dense, medium, or open, depending on the percentage of abrasive or pores. Wheels of medium structure are best for heavy cuts and for soft, ductile materials that are easily penetrated and require good chip clearance.

The size of the abrasive grain is determined by the size of the mesh screen through which the abrasive particles will fall. The five types of bonding materials that are used to hold the particles together are vitrified, resinoid, rubber, shellac, and silicate.

Grinding wheel manufacturers have standardized 9 shapes and 12 faces for grinding wheels. These wheels are made in a wide range of sizes to do most grinding jobs. Each wheel should be checked for balance before mounting. Out-of-balance wheels set up excessive vibrations, causing chatter marks on ground surfaces.

The bond in a grinding wheel holds the grains together and supports them while they cut. The greater the amount of bond with respect to the abrasive, the heavier the coating around the abrasive grains and the harder the wheel. The abrasive itself is extremely hard in all grinding wheels, and the terms *hard* and *soft* refer to the strength of the bond in the wheel.

A standard system for marking grinding wheels is used to designate the kind of abrasive, grain size, grade structure, bonding process, and the manufacturer's markings.

Review Questions

1. Name the two abrasives most commonly used for making grinding wheels.
2. How is the grain size of an abrasive measured?
3. What is meant by the bond of a grinding wheel?
4. Name several common shapes of grinding wheels.
5. What type of abrasive is used to grind nonferrous and nonmetallic materials?
6. What is the most common natural grindstone?

7. How many shapes and faces are there for grinding wheels?

8. List at least six grinding wheel standard shapes.

9. Sketch at least four grinding wheel faces.

10. What do out-of-round grinding wheels do?

11. What does *hard* and *soft* refer to in terms of grinding wheels?

12. What are the three subdivisions of bonds used in organic grinding wheels?

13. List the six parts of the code used to mark grinding wheels.

14. What determines the kind of abrasive selected for a job?

15. Why is grinding wheel speed important?

Chapter 6
Cutting Fluids

Cutting fluids are primarily used in machining operations to reduce temperature and adhesion between the chip and tool. They also serve to keep the workpiece cool to avoid thermal expansion and provide easier handling. When a lubricant is part of the cutting fluid, it provides a rust-proof layer to the finished work surface. Cutting fluids are useful in clearing chips away from the machining area.

Coolant

A *coolant* is an agent (usually in liquid form) whose sole function is to absorb heat from the work and the cutting tool. Water has the highest cooling effect of any cutting fluid. It may be used on materials that are tough or abrasive, but have a great frictional effect and generate much heat in cutting. Typical materials of this nature are rubber tires and celluloid.

Lubricants

A cutting fluid with the additional property that enables it to act as a lubricant is called a *cooling lubricant*. These materials usually consist of both cooling and lubricating agents, such as soluble oil or glycerin mixed in proper proportions. Cooling lubricants are used where cutting materials generate excessive heat and are, to a limited degree, tough and abrasive.

Application of Cutting Fluids

A stream of soluble oil is frequently used to increase the cutting capacity of an abrasive wheel by preventing it from glazing over, and by carrying off the heat generated by the friction of the wheel on the work. It is important that the coolant be properly applied to the work. A large stream of fluid at slow velocity is preferable to a small stream at high velocity. The cutting fluid should make contact with the work at the exact spot where the cutting action takes place—not above or to one side of the cutting tool (Figure 6-1).

Limited amounts of cutting fluid may be applied with an oilcan. A small paintbrush can also be used to apply limited amounts of coolant in many instances.

On machine tools, cutting fluids keep heat from softening and ruining the cutting edge of the cutting tool. A cutting fluid cools the

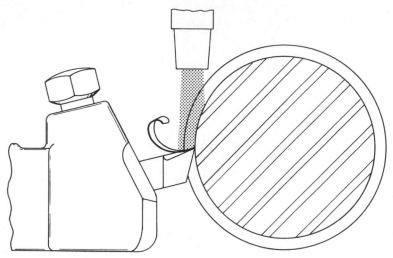

Figure 6-1 Cutting fluid should make contact with the work where the cutting action takes place. *(Courtesy South Bend Lathe, Inc.)*

cutting tool and makes it cut more easily and smoothly. The fluid also tends to wash away chips, prevents undue friction, and permits faster cutting speeds.

Production machine tools and many general-purpose machines are equipped with oil pans, pumps, and reservoirs, which circulate the cutting fluid to the point of cutting. Spray-mist coolant systems use a water-base fluid that is supplied to the cutting tool by compressed air. The compressed air atomizes the cutting fluid providing considerable cooling but little lubrication.

Types of Cutting Fluids

The characteristics of the most commonly used cutting fluids are as follows:

- *Lard oil*—This is one of the oldest and best cutting oils. It provides excellent lubrication, increases tool life, produces a smooth finish on the work, and prevents rust. This coolant is especially good for cutting screw threads, drilling deep holes, and reaming. Its industrial application is limited because it is very expensive, tends to become rancid, and can cause skin irritation.

- *Mineral-lard oil mixtures*—These mixtures are used in various proportions of lard oil and petroleum-base mineral oils in place of lard oil alone because they are more fluid, less expensive, and almost as effective.

- *Mineral oils*—Petroleum-base oils are compounded with chemicals to improve their lubricating and anti-welding qualities. They are also less expensive than lard oil and mineral oil mixtures.

- *Water-soluble oils*—Their use is limited to rough turning operations. Although they carry away heat better than lard oil or mineral oil, their lubricating qualities are poor. They are treated so that they may be mixed with water to form an emulsion and provide an excellent low-cost coolant.

- *Chemical cutting fluids*—These compounds are mixed with water and generally do not contain any petroleum products. The chemicals used provide excellent lubricating and anti-welding properties. These solutions are gradually replacing other cutting fluids in a variety of applications, including several machining operations and surface grinding.

- *Low-viscosity oils*—Oils such as kerosene are used for cutting tough nonferrous metals and alloys (materials such as the bronzes, certain aluminum alloys, and alloys containing a very small percentage of iron such as Monel metal). The heavier, more sulfurnated oils are used for sawing tough steels and ferrous alloys because these oils are better able to withstand extreme pressure and abrading actions.

Solid Lubricants

Grease-type lubricants are used to lubricate the sides and back edge of a band saw and the guide bearings, particularly where high velocity is involved. These lubricants are often compounded with graphite and supplied in tubes. On some tubes, a threaded nozzle permits the tube to be screwed into a grease fitting associated with a saw-guide mounting. Contents of the tube can either trickle through or be squeezed through apertures in the mounting. The tube can remain fastened to the mounting to lubricate the blade until the tube is empty.

The wax or soap lubricants are usually compounded of petroleum substances; many are impregnated with high film-strength lubricants. This lubricant is normally supplied in stick form (such as a cardboard tube) from which it is pushed out for use. A good solid lubricant is recommended for sawing metals, hardwoods, and other material where the use of surface lubricant is preferable to a circulating type.

Cutting oils are usually dripped onto the cutting edge of the tool at a rate of 6 to 40 drops of oil per minute, depending on the material being cut and the speed at which it is cut. Table 6-1 gives the cutting speeds and lubricants for both drilling and turning the various materials. Table 6-2 shows cutting fluid recommendations for sawing various materials.

Table 6-1   Cutting Speeds and Fluids for Turning and Drilling with High-Speed Steel Cutting Tools

Material	Turning Speed in Ft per Minute	Cutting Fluid	Drilling Speed in Ft per Minute	Cutting Fluid
Aluminum	300–400	Kerosene	200–330	Kerosene
Brass	300–700	Dry	200–500	Dry
Bronze	300–700	Compound	200–500	Compound
Cast iron	50–110	Dry	100–165	Dry
Copper	300–700	Compound	200–500	Compound
Duralumin	275–400	Compound	250–375	Compound
Fiber	200–300	Dry	175–275	Dry
Machine steel	115–225	Compound	80–120	Compound
Malleable iron	80–130	Dry or compound	80–100	Dry or compound
Monel metal	100–125	Compound	40–55	Sulfur base
Plastics, hot-set molded	200–600	Dry	75–300	Dry
Rubber, hard	200–300	Dry	175–275	Dry
Stainless steel	100–150	Sulfur base	30–45	Sulfur base
Tool steel	70–130	Compound	45–65	Sulfur base

Courtesy South Bend Lathe, Inc.

Table 6-2   Cutting Fluid Recommendations for Sawing Various Materials

Material	Lubricant or Coolant
Aluminum; sheet, rod, bar, press forgings (2S, 3S, 4S, 11S, 17S)	Coolant-lubricant
Autobestos and Raybestos (brake lining)	Coolant-lubricant
Commercial brass sheet (SAE No. 70) quarter hard to extra spring	None

Table 6-2   (continued)

Material	Lubricant or Coolant
Atlas 90 and Auromet 55 aluminum bronze	Light cutting oil
Felt (woolen or cotton)	None
Granite (Igneous)	Coolant-lubricant
Optical glass	Coolant-lubricant
Magnesium die castings SAE 501	Mineral oil
Bakelite (no filler)	None
Polystyrene	Coolant-lubricant
Buna rubber	Glycerine and water
SAE (Carbon steel) 1006, 1010, 1015, 1020, 1025, 1030	Heavy cutting oil
SAE (Chromium-vanadium steel) 6130, 6135, 6140, 6150	Heavy cutting oil
SAE (Stainless) 30–303	Coolant-lubricant
Kestos steel	Heavy cutting oil
Meehanite	None

It is usually more convenient to use the fluid lubricants because they can be circulated automatically with a resultant saving in time. The fluid coolants and lubricants can be applied either in a stream or by dripping or spraying.

The solid lubricants are usually applied directly to the surface. For example, in band filing of nonferrous metals, the file segment may become loaded and difficult to clean. The solid wax lubricants may be applied directly to the filing surface to prevent loading.

Summary

Water is the only agent whose sole function is cooling. Coolants are used as cooling agents to carry off heat generated by friction of the grinding wheel on the work.

Other materials are used primarily for their cooling or lubricating properties. These include cutting oils, semisolid greases, and solid lubricants. Not only are cooling lubricants important in carrying away heat, and so on, but they are also important in properly cooling the cutting edge of the tool and the grinding stone.

Cutting fluids are primarily used in machining operations to reduce temperature and adhesion between the chip and tool. They are also used to keep the piece cool and to avoid thermal expansion and to make for easier handling.

Cooling lubricants are used where cutting materials generate excessive heat and are, to a limited degree, tough and abrasive.

Various types of cutting fluids are available. They are lard oil, mineral-lard oil mixtures, mineral oils, water-soluble oils, chemical cutting fluids, and low-viscosity oils.

Wax and soap are also used in cooling and lubricating. The wax or soap lubricants are usually compounded of petroleum substances and some are impregnated with high film-strength lubricants. These are usually supplied in a stick form with a cardboard cover and are used by pushing them out of the cardboard cylinder.

However, it is usually more convenient to use fluid lubricants because they can be circulated automatically with a resultant saving in time. Solid lubricants are usually applied directly to the filing surface to prevent loading.

Review Questions

1. What is the purpose of coolants?
2. Name a few materials used primarily for cooling.
3. Name a few solid coolants used.
4. Define the term *coolant*.
5. A cutting fluid with the additional property that enables it to act as a _____ is called a cooling lubricant.
6. Why are coolants necessary?
7. What is lard oil?
8. What are water-soluble oils?
9. How are wax and soap used as lubricants?
10. Where are low-viscosity oils used?

Chapter 7

Cutting Tools

All operators of machine tools should have a basic knowledge of the cutting action of the cutting tools. This understanding is necessary for the cutting tool to be ground properly and applied to the work correctly.

Action of Cutting Tools

Cutting tools employ a wedging action. All the power used in cutting metal is ultimately expended in heat. A tool that has been used on heavy cuts has a small ridge of metal directly over the cutting edge. This bit of metal is much harder than the metal being cut, and is almost welded to the edge of the tool, indicating that an immense amount of heat and pressure was developed.

The fineness or sharpness of the edge (the angle of the two sides of the tool that make the edge) depends on the class of work (roughing or finishing) and on the metal being cut. It is not necessary to hone the edge of the cutting tool for heavy roughing cuts in steel. A fine edge lasts for only a few feet of cutting, then rounds off to a more solid edge and remains in approximately the same condition until the tool breaks down.

In high-speed production work, coolants help absorb the heat from the cutting edge of the tool. A steady stream of cutting compound should be directed at the point of the cutting tool so that it spreads and covers both the cutting tool and the work. A large pan collects the cutting compound, carries it to a setting tank, and then to a pump.

Materials

There are several different materials used to make cutting tools or cutter bits. To machine metal accurately and efficiently, it is necessary to have the proper lathe tool ground for the particular kind of metal being machined, with a keen, well-supported cutting edge. Following are some of the materials used to make cutting tools:

- *Carbon tool steel*—These cutting tools are less expensive and can be used on some types of metal successfully.
- *High-speed steel*—Cutter bits made of high-speed steel are the most popular cutting tools. They will withstand higher cutting

speeds than carbon steel cutter bits. High-speed tools contain tungsten, chromium, vanadium, and carbon.

- *Stellite*—These cutter bits will withstand higher cutting speeds than high-speed steel cutter bits. Stellite is a nonmagnetic alloy that is harder than common high-speed steel. The tool will not lose its temper, even though heated red hot from the friction generated by taking the cut.

  Stellite is more brittle than high-speed steel and requires less clearance, or just enough clearance to permit the tool to cut freely, to prevent breaking or chipping. Stellite is also used for machining hard steel, cast iron, bronze, and so on.

- *Carbide*—Tips of cutting tools are made of carbide for manufacturing operations where maximum cutting speeds are desired.

- *Tungsten carbide*—These cutting tools are efficient for machining cast iron, alloyed cast iron, bronze, copper, brass, aluminum, Babbit, and abrasive nonmetallic materials (such as hard rubber, fiber, and plastics). These cutter bits are so hard that they cannot be ground satisfactorily on an ordinary grinding wheel.

- *Tantalum carbide*—The term tantalum is applied to a combination of tungsten carbide and tantalum carbide. These cutting tools are similar to tungsten carbide tools, but are used mostly for machining steel.

- *Titanium carbide*—This is a term applied to a combination of tungsten carbide and titanium carbide. Titanium carbide is interchangeable with tantalum carbide in its uses.

Shapes and Uses of Cutting Tools

Following are nine of the most popular shapes of lathe cutter bits and their applications (see Figure 7-1):

- *Left-hand turning tool*—The opposite of the right-hand turning tool, this tool is used to machine work from left to right (Figure 7-1A).

- *Round-nose turning tool*—The round-nose tool is a convenient tool for turning in either direction and for reducing the diameter of a shaft in the center (Figure 7-1B).

- *Right-hand turning tool*—This tool is the most common type of cutting tool for general lathe work. It is used for machining work in which the cutting tool travels from right to left (Figure 7-1C).

- *Left-hand facing tool*—This tool is the opposite of the right-hand tool, and is used for facing the left-hand side of the work (Figure 7-1D).

- *Threading tool*—The point of this tool is ground to an included angle of 60°. It is used to cut screw threads (Figure 7-1E).

- *Right-hand facing tool*—This cutting tool is used for facing the ends of shafts and for machining work on the right side of the shoulder (Figure 7-1F).

- *Cutoff tool*—This tool is used for various classes of work that cannot be accomplished with a regular turning tool (Figure 7-1G).

- *Boring tool*—This tool is ground the same as the left-hand turning tool, except that the front clearance angle is slightly greater, so that the heel of the tool will not rub in the hole of the work (Figure 7-1H).

- *Inside-threading tool*—This tool is ground the same as the screw-threading tool, except that the front clearance angle must be greater (Figure 7-1I).

Heavy forged turning tools were formerly used, but in more recent years they have been replaced by toolholders and smaller tools made of high-speed steel, known as *tool bits*. These tool bits are generally used in a standard toolholder that presents them to the work at an angle of 15° to 20° with the horizontal axis of the work. This angle of inclination is important and must always be considered in the sharpening of every tool bit that is to be used in a toolholder.

Terms Related to Cutting Tools

Following are some terms and definitions related to cutting tools:

- *Base*—Surface of the shank that bears against the support and takes the tangential pressure of the cut.

- *Chip breaker*—An irregularity in the face of a tool or a separate piece attached to the tool or holder to break the chips into short sections.

- *Cutting edge*—Portion of the face edge along which the chip is separated from the work.

- *Face*—Surface on which the chip slides as it is cut from the work.

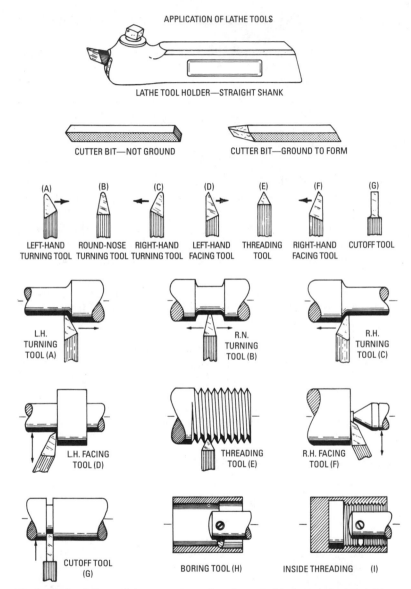

APPLICATION OF LATHE TOOLS

LATHE TOOL HOLDER—STRAIGHT SHANK

CUTTER BIT—NOT GROUND

CUTTER BIT—GROUND TO FORM

(A) LEFT-HAND TURNING TOOL

(B) ROUND-NOSE TURNING TOOL

(C) RIGHT-HAND TURNING TOOL

(D) LEFT-HAND FACING TOOL

(E) THREADING TOOL

(F) RIGHT-HAND FACING TOOL

(G) CUTOFF TOOL

L.H. TURNING TOOL (A)

R.N. TURNING TOOL (B)

R.H. TURNING TOOL (C)

L.H. FACING TOOL (D)

THREADING TOOL (E)

R.H. FACING TOOL (F)

CUTOFF TOOL (G)

BORING TOOL (H)

INSIDE THREADING (I)

Figure 7-1 Nine of the most popular shapes of lathe tool cutter bits and their application. *(Courtesy South Bend Lathe, Inc.)*

- *Flank*—Surface or surfaces below and adjacent to the cutting edge.
- *Flat*—Straight portion of the cutting edge intended to produce a smooth machined surface.
- *Heel*—Edge between the base and the flank immediately below the face.
- *Neck*—An extension of reduced sectional area of the shank.
- *Nose*—A curved face edge.
- *Shank*—Portion of the tool back of the face that is supported in the tool post.

Cutting Tool Angles

The cutting end of the cutting tool is adapted to its cutting requirements by grinding its sides and edges at various angles. Since the cutting tool is more or less tilted in the toolholder, the angles are classed as either *tool angles* or *working angles*.

Tool angles are those angles applied to the tool itself, whereas working angles are those angles between the tool itself and the work. Working angles depend not only on the shape of the tool, but also on its angular position with respect to the work.

Tool Angles

The cutting tool itself must have both the rake angle and the relief angle ground at the proper angle, depending on the cutting requirements (Figure 7-2).

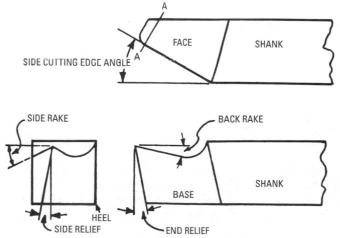

Figure 7-2   Cutting tool rake and clearance angles.

Back Rake

All cutting tools employ a wedging action. The differences are in the angle of the two sides of the tool bit that form the cutting edge and the manner in which the cutting tool is applied to the work.

The edge of a pocketknife would be ruined in trying to cut a nail, even though the metal in the knife is much harder than that in the nail. A cold chisel, however, shows no sign of damage in cutting the same nail, although the chisel is usually a poorer grade of steel than the knife. Obviously, the difference is in the angle of the cutting edge.

Back rake is usually provided for in the toolholder by setting the tool at an angle. This angle may vary from 15° to 20°. Back rake may be varied for the material being turned by adjusting the tool-holder in the post through the rocker.

The inclination of the face of a tool is to or from the base. If it inclines toward the base, the rake angle is *positive*. If it inclines away from the base, the rake angle is *negative* (Figure 7-3).

The cutting angle should be as large as possible for maximum strength at the edge and to carry heat away from the cutting edge. On the other hand, the larger the cutting angle, the more power is

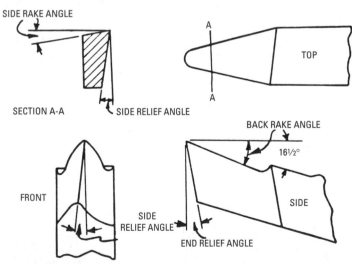

Figure 7-3   A round-nose, right-hand cutting tool, suitable for roughing and general-purpose turning. By increasing the front rake and by using no side rake, it can be used for either right-hand or left-hand turning.

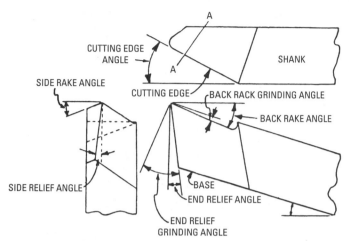

Figure 7-4 Working angles of the cutting tool, that is, the actual angles with the cutting tool tilted either in the tool post or in the toolholder.

required to force it into the work. Thus, a compromise is necessary to arrive at the best rake angles (Figure 7-4).

Top rake should be less for softer material, as there is a tendency for the tool to dig in if the rake angle is too great. There should be no top rake angle for turning bronze; the cutting edge of the tool should be almost horizontal. A negative rake is used for turning soft copper, Babbitt, and some die casting alloys (Figure 7-5).

Side Rake
The side rake angle also varies with the material being machined. Side rake is the angle between the face of a tool and a line parallel to the base. The cutting tool will not cut without side rake, and this angle relieves excessive strain on the feed mechanism. The proper side rake angle varies from 6° for soft material to 15° for steel (Figure 7-6 and Figure 7-7).

End Relief
This is the angle between the flank and a line from the cutting edge perpendicular to the plane of the base. End relief depends somewhat on the diameter of the work to be turned. If a tool were ground square, without any front clearance, it would not cut because the material being turned would rub on the cutting tool just below the cutting edge (Figure 7-8).

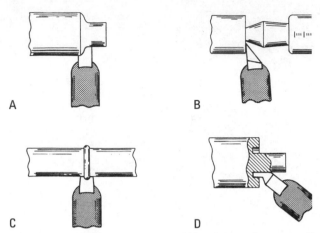

A   B

C   D

Figure 7-5 Showing some of the special form cutting tools and their applications. These are called form tools to distinguish them from regular types. In form cutting tools such as A and D, side rake is not used. Front rake, however, should be used, except when turning brass. Form cutting tools wider than 1/8 inch should not be used on steel. Form cutting tools as wide as 1/2 inch can be used on brass, aluminum, and similar metals.

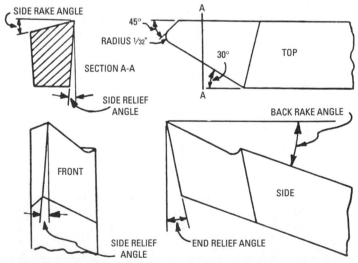

Figure 7-6 A right-hand turning tool, showing the tool angles.

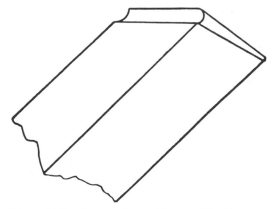

Figure 7-7 A special roughing tool for heavy-duty work in steel. A large side-rake angle can be obtained without unduly weakening the tool by grinding a groove along the edge of the tool, instead of grinding away the top of the tool at an angle. The radius of the groove largely determines the diameter of the chip curl. This entire groove should be honed smoothly because the entire face of the chip bears on the surface of this groove, and any roughness increases friction and heat.

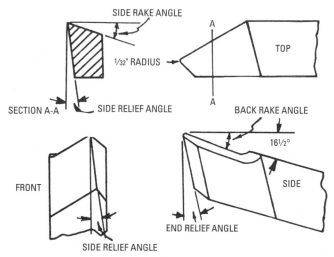

Figure 7-8 A left-hand turning tool, showing the various tool angles.

End relief should be less for small diameters than for large diameters, and may range from 8° to 15°. Do not grind more front clearance than is necessary, as this takes away support from the cutting edge of the tool.

Side Relief

This is the angle between the side of a tool and a line from the face edge perpendicular to the plane of the base (Figure 7-9 and Figure 7-10). In turning, the relief angle allows the part of the tool bit directly under the cutting edge to clear the work while taking a chip.

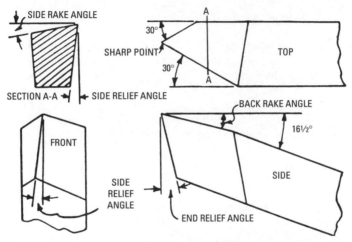

Figure 7-9 A right-hand facing tool, showing the various tool angles.

Too much relief weakens the cutting edge for clearing the work while taking a chip. The high pressure exerted downward on the tool bit demands that the relief be as small as possible and still allow the tool bit to cut properly. Whenever the tool bit digs into the work, or refuses to cut unless forced, the relief of the tool bit should be checked. Digging in occurs most often during facing and threading operations. For light turning operations, it is usually better to allow slightly more than enough relief rather than risk too little relief.

Working Angles

The angle between the tool and the work depends not only on the cutting tool angle, but also on the position of the cutting tool in the toolholder.

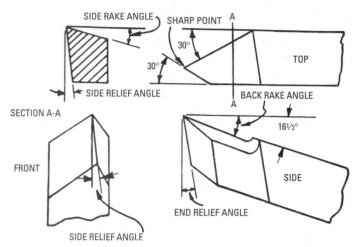

Figure 7-10 A left-hand facing tool, showing the various tool angles.

Setting Angle
The setting angle is the angle the cutting tool axis makes with the work axis when set.

True Top Rake Angle
The top rake angle plus the setting angle gives the true top rake angle (Figure 7-11).

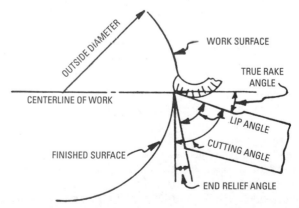

Figure 7-11 Angles of the tool bit in relationship to the work.

Side Cutting Edge Angle
The side cutting edge angle is the angle between the face of the cutting tool and tangent to the work at the point of cutting (see Figure 7-2). Here, again, the cutting angle can be varied to suit the material being turned by adjusting the toolholder in the tool post.

Angle of Keenness
The angle of keenness is the included angle of the tool between the face of the tool and the ground flank adjacent to the side cutting edge.

Table 7-1 shows rake and clearance angles measured in degrees.

Table 7-1   Rake and Clearance Angles in Degrees

Material	Top Rake	Side Rake	Front Clearance	Side Clearance
Cast iron	5	12	8	10
Stainless steel	16½	10	10	12
Copper	16½	20	12	14
Brass and softer copper alloys	0	0	8	10
Harder copper alloys	10	0–2	12	10
Hard bronze	0	0–2	8	10
Aluminum	35	15	8	12
Monel metal and nickel	8	14	13	15
Phenol plastics	0	0	8	12
Various base plastics	0–5	0	10	14
Formica gear plastic	16½	10	10	15
Fiber	0	0	12	15
Hard rubber	0 or −5	0	15	20

High-Speed Steel Lathe Tools
Depending on the job, the lip and clearance angles ground on high-speed tools vary in size. The tool is weakened by a clearance angle that is too great, while too little clearance may permit the tool to rub against the work. The lip, which is ground on top of the cutting tool parallel to the cutting edge, causes the chip to curl, and controls the chip. The lip should be ground ⅛ to ¼ inch wide and about ⅙ inch deep. Coarse feeds and deep cuts require a large lip (Figure 7-12 and Figure 7-13).

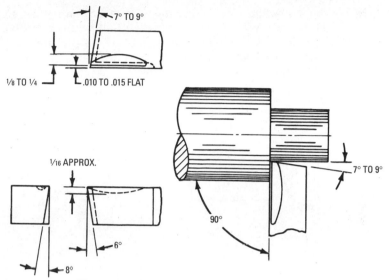

Figure 7-12 Turning tool used in a four-way turret tool post, showing lip and clearance angles.

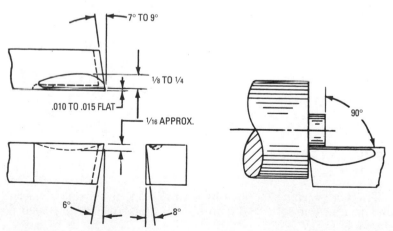

Figure 7-13 Facing tool used in a four-way turret tool post, showing lip and clearance angles.

The tool bit should be mounted in the toolholder at an angle of approximately 15½°. A 21° angle must be ground on the front of the tool bit (Figure 7-14 and Figure 7-15). A finish-turning and facing tool may be used for finish turning an outside diameter and facing in one setting (Figure 7-16).

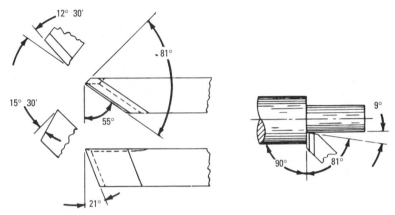

Figure 7-14   Turning tool bit, as used in toolholder.

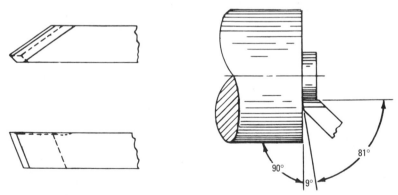

Figure 7-15   Facing tool bit, as used in toolholder.
(Courtesy Cincinnati Milacron Company.)

A *corner necking tool* can be used to neck an outside diameter and face in one setting (Figure 7-17). An *offset necking tool* is shown in Figure 7-18.

The *counterbore* (or *corner tool*) can be used for boring steps in soft chuck jaws, as well as for counterboring. More clearance must

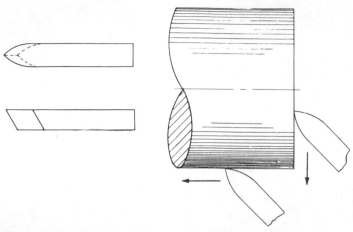

Figure 7-16   Finish turning an outside diameter and facing in one setting.

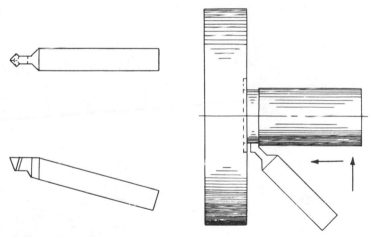

Figure 7-17   Corner necking tool used to neck an outside diameter and to face in one setting.

be ground on the tool when smaller holes are to be counterbored (Figure 7-19).

A *cutoff tool* is shown in Figure 7-20. The angles on the cutoff tool will vary, depending on the width of the tool.

Boring tools are held in special toolholders. The boring tool is designed primarily for small work. These tools are sharpened in much the same manner as other lathe tools (Figure 7-21).

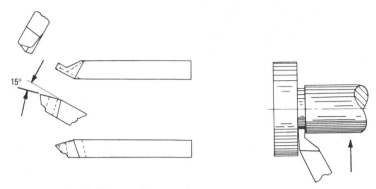

Figure 7-18 Offset necking tool. *(Courtesy Cincinnati Milacron Company.)*

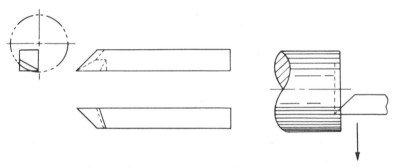

Figure 7-19 Counterbore or corner tool. *(Courtesy Cincinnati Milacron Company.)*

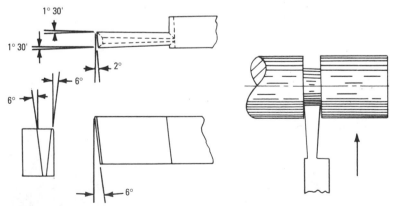

Figure 7-20 Cutoff tool. The angles will vary depending on the width of the tool. *(Courtesy Cincinnati Milacron Company.)*

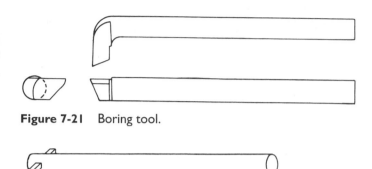

Figure 7-21   Boring tool.

Figure 7-22   Boring bar with tool bit.

Boring bars are designed for larger work. The tool bit is inserted into a slot in the boring bar and held by a set screw or other means. The bars are made with the slot at a different angle to the centerline of the bar (45°, 60°, and 90°) for different types of work (Figure 7-22).

Summary

Cutting tools used in shop work are generally wedge-shaped. A tool that has been used on heavy cuts develops a small ridge of metal directly over the cutting edge. This bit of metal is much harder than the metal being cut and is almost welded to the edge of the cutting tool, which indicates the amount of heat and pressure that has been developed. The fineness of the edge depends on the class of work (either roughing or finishing) and on the metal being cut. In high-speed production work, coolants help absorb the heat from the cutting edge of the tool. A steady stream of cutting compound should be directed at the point of the cutting tool so that it spreads and covers both the cutting tool and the work.

Various materials are used in making cutting tools or bits. Some of the materials used to make cutting tools are carbon steel, high-speed steel, stellite (a nonmagnetic alloy that is harder than common high-speed steel), carbide, tungsten, tantalum, and titanium. Nine of the most popular shaped lathe cutter bits are left-hand and right-hand turning tools, left-hand and right-hand facing tools, round-nose turning tools, threading tools, cutoff tools, boring tools, and inside-threading tools.

The angle between the tool and the work depends, not only on the cutting tool angle, but also on the position of the cutting tool in the tool holder. Rake and clearance angles are such that they require a chart to present the various angles needed for the various metals that can be cut.

Boring tools are held in special tool holders. The boring tool is designed primarily for small work. They are sharpened much the same as other lathe tools.

Review Questions

1. Why are coolants used with cutting tools?

2. Name at least four cutting tools used in machine shops.

3. What type of material is used to make cutting tools?

4. What is a tool angle or working angle?

5. What happens to the cutting tool edge after it has been used for a few minutes?

6. What is the purpose of a settling tank?

7. How is the round-nose turning tool used?

8. What is a tool bit?

9. What is a chip breaker?

10. Why is the rake angle of a cutting tool important?

11. What is meant by side rake?

12. Where do you use a corner necking tool?

13. Where is the counterbore, or corner tool, used?

14. What is a boring bar used for?

15. How is the cutoff tool used to cut off material?

16. Identify the following terms:

  a. Flank

  b. Flat

  c. Nose

  d. Neck

  e. Shank

  f. Heel

  g. Face

  h. Base

  i. Chip breaker

17. What is back rake?

18. What is top rake?

19. What is meant by "end relief"?

20. What is the angle of keenness?

Chapter 8

Cutter and Tool Grinders

A variety of metal-cutting applications can be performed to close tolerances with modern precision tools. Unless the machine has been provided with suitably designed and prepared cutting tools, close tolerances are impossible. The machine, no matter how precise, can be no better than the tool with which it is equipped.

Importance of Tool Sharpening

If a cutter is used after it becomes dull, it deteriorates rapidly. Very little stock needs to be ground off if the cutter is sharpened at the proper time. Only trained operators with suitable equipment, using properly specified grinding wheels, should sharpen lathe and planer tools (Figure 8-1). Proper sharpening also extends the life and increases the efficiency of twist drills.

Figure 8-1 Tool-sharpening machine. Direct-reading scales and dials permit grinding different angles, plus the tip radius, in one operation. An extra wheel and freehand grinding attachment are available on the left-hand side of the machine for sharpening odd-shaped tools, or for roughing shapes that will be precision finished on the right-hand side of the machine. *(Courtesy Heald Machine Company.)*

Cutter and Tool Sharpening

The universal cutter and tool grinding machines meet practically all tool-room requirements. Cutter grinding machines may vary from simple machines for sharpening multitooth cutters to universal grinding machines.

Lathe and Planer Tools

Either the offhand method or the machine method may be used to sharpen lathe and planer tools. The skill of the operator determines the accuracy of a tool ground by the offhand method (Figure 8-2). Precise rake and relief angles can be reproduced by means of dial settings when tools are sharpened on a machine (Figure 8-3). Localized overheating may crack the cutting tool unless it is kept constantly in motion across the grinding-wheel face.

Figure 8-2   Freehand grinding on the left-hand side of the cutting tool sharpening machine. *(Courtesy Heald Machine Company.)*

Twist-Drill Sharpening

Machine grinding is a more accurate method than grinding by hand when sharpening twist drills. A twist drill usually cuts faster, lasts longer, and produces more accurate holes when ground on a machine, as compared to grinding by hand.

The suggestion to an old-time mechanic that he use a machine to sharpen a twist drill would probably elicit a scornful retort such as,

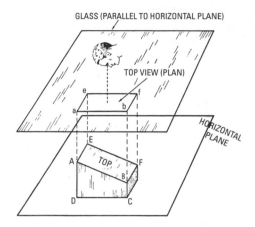

GLASS (PARALLEL TO HORIZONTAL PLANE)

TOP VIEW (PLAN)

HORIZONTAL PLANE

TOP

Figure 8-3 Mechanical sharpening on the right-hand side of the cutting tool sharpening machine.

(Courtesy Heald Machine Company.)

"I can grind a drill as well by hand as with a machine." Undoubtedly, a skilled mechanic can produce a satisfactory drill point by exercising great care and taking plenty of time, because this is not an impossible task. However, all drill operators are not skilled mechanics, and in this advanced age of mechanical knowledge, a modern up-to-date machine shop will not rely on a skilled mechanic for this precise operation. It will not tolerate the use of hand methods when there is an inexpensive machine available to do the job more accurately and quickly.

If we consider the essential points in connection with sharpening twist drills, the fallacy of any mechanic attempting to sharpen a drill by hand can be seen readily (see Figures 8-4 through 8-6). Drill sharpening requirements are as follows:

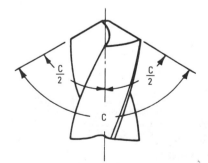

Figure 8-4 The two cutting edges (lips) should be equal in length, and should form equal angles with the axis of the drill. Angle C should be 135° for drilling hard or alloy steels. For drilling soft materials and for general purposes, angle C should be 118°.

(Courtesy National Twist Drill & Tool Company.)

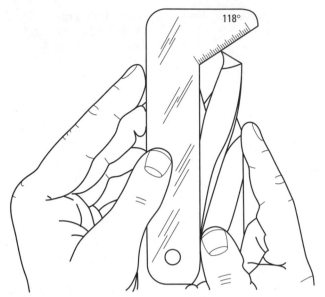

Figure 8-5 Using a drill-grinding gage to check the drill bit angle.

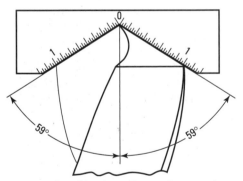

Figure 8-6 Checking the angle to make sure they are equal. Otherwise, one side will do all of the cutting.

1. The cutting lip of the twist drill should form equal angles with the axis of the drill. (The commercial standard angle is 59°.) A machinist must possess a good eye, indeed, to gage an angle of 50° with the eye. Drill manufacturers have found 59° to be the best angle for all-around work.

If one lip of a twist drill is ground at 60° and the other at 59°, it is easy to see that the 60° lip will do all of the cutting. Half-speed drilling, half the length of service between grindings, conical cone, double wear, waste of the drill, and so on, are all results of unequal lip angles (Figure 8-7).

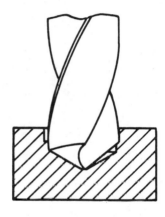

Figure 8-7 A twist drill with the lips ground at unequal angles with the axis of the drill can be the cause of an oversized hole. Unequal angles also result in unnecessary breakage and cause the drill to dull quickly.

(Courtesy National Twist Drill & Tool Company.)

2. The cutting lips should be of equal length. Even though the lip angles are equal, a twist drill with cutting lips that are unequal in length makes oversize holes and causes strain on both the twist drill and the drill press (Figure 8-8). When one lip is only slightly longer, that lip does all of the cutting for that extra length. This means rapid wear, frequent regrinding, and slower drilling speed (Figure 8-9).

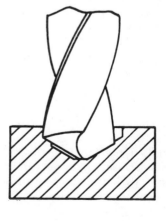

Figure 8-8 This shows the result of grinding the drill with equal angles, but unequal lips.

(Courtesy National Twist Drill & Tool Company.)

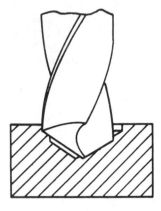

Figure 8-9 This shows the result of grinding the drill with lips of unequal angles and with lips of unequal lengths.
(Courtesy National Twist Drill & Tool Company.)

3. The lip *clearance angle* (or lip *relief angle*) should gradually increase as it approaches the center of the twist drill (Figure 8-10). If the twist drill is to penetrate the work and its edge is to cut, the surface behind the cutting edge must be ground away at an angle, giving what is termed *clearance*. If there is no clearance angle, the tool will ride along the surface without entering the work. The clearance angle determines the effectiveness of the twist drill and its length of life (Figure 8-11).

Figure 8-10 The lip relief angle A should vary according to the material to be drilled and the diameter of the drill. Lesser relief angles are required for hard and tough materials than for soft, free-machining materials.
(Courtesy National Twist Drill & Tool Company.)

In order to grade the clearance properly along the drill lip from point to periphery, and in order to curve the back side of the cutting edge so that maximum endurance and strength consistent with free cutting are preserved at all points, it is necessary that every portion of the cutting lip, while being ground, rock against the grinding wheel in a path similar to that in which the cutting lip travels when at work. If, while at work, those portions of the cutting lip nearer the point travel shorter paths and make smaller circles than the portions nearer the outer corner, this similar condition should exist while the twist drill is being ground. Table 8-1 shows the suggested lip relief angle at the periphery.

The webs of twist drills increase in thickness toward the shank of the drill, in order to increase strength and rigidity. The proper web thickness should be maintained as the drills are sharpened. Thickness of the web may be further reduced for some applications to not less than 50 percent of the original web thickness (Figure 8-12).

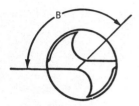

Figure 8-11 The chisel point angle B increases or decreases with the relief angle, but it should range from 115° to 135°. *(Courtesy National Twist Drill & Tool Company.)*

Table 8-1   Suggested Lip Relief Angle at the Periphery

Drill Diameters	For General Purpose	Hard and Tough Materials	Soft and Free-Machining Materials
No. 80–No. 61	24°	20°	26°
No. 60–No. 41	21°	18°	24°
No. 40–No. 31	18°	16°	22°
No. 30–1/4"	16°	14°	20°
F to 11/32"	14°	12°	18°
S to 1/2"	12°	10°	16°
33/64"–3/4"	10°	8°	14°
49/64" & lgr.	8°	7°	12°

Courtesy National Twist Drill & Tool Company.

Automatic drill grinders are designed and constructed in accordance with geometric principles. These machines automatically locate the cone axis in relationship to the drill axis for all twist drill sizes within the capacity of the machine; they also automatically grind the correct lip clearance angle and the proper center angle.

Regrinding Tap Drills
Usually the tap drill cutting edges become dull first, and it may be necessary to grind only that portion of the tap drill. Special

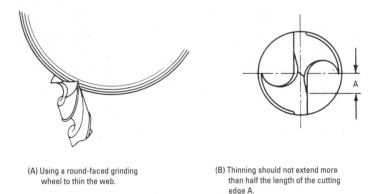

(A) Using a round-faced grinding wheel to thin the web.

(B) Thinning should not extend more than half the length of the cutting edge A.

Figure 8-12 Thinning the web at the point of the drill. *(Courtesy National Twist Drill & Tool).*

machines are available for sharpening taps accurately, ensuring a uniform chamfer or taper, and correct, uniform eccentric relief. Figure 8-13 and Figure 8-14 show drill bit sharpeners.

Mounted Points and Wheels
Miscellaneous offhand grinding operations on all kinds of blanking and drop-forging dies are performed with a wide variety of mounted points and mounted wheels. They are necessary for shaping and finishing steel dies and molds that are used in the plastic and allied tool-and-die industries. The mounted wheels are usually driven by light, flexible-shaft, air- or motor-driven portable grinders. Surplus metal can be removed in much less time with this equipment than by scraping or filing, and a better finish may be obtained.

Cutter-Sharpening Machines
The machines designed to grind cutters may vary from single type for grinding some of the multitooth cutters, to the universal cutter and tool grinding machines, which have a range for practically all tool-room requirements. Many of the cutter and tool grinders have special attachments for grinding the various milling machine cutters (Figures 8-15 through 8-19).

Plain Milling Cutters
The cutter may be mounted on an arbor and the tooth rest adjusted to give the desired relief angle. The tooth rest should bear against the tooth to be ground wherever possible. The cutter should be moved slowly toward the wheel until sounds or sparks indicate

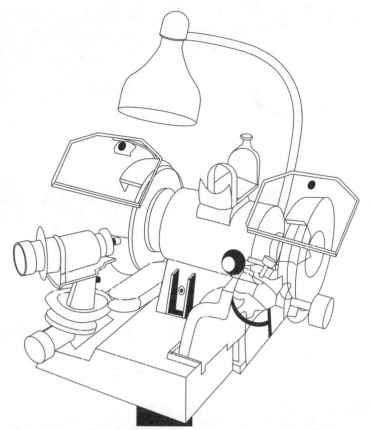

Figure 8-13 Drill bit sharpener.

contact. Holding the cutter against the tooth rest with one hand, the cutter is traversed with the other hand across the wheel face with a steady motion, either by moving the table or by sliding the cutter on a cutter bar. The cuts should not exceed 0.001 inch per pass on roughing and should be reduced to 0.0005 inch on finishing passes.

Side Milling Cutters
Cutter teeth of the side mill type are ground on the outside diameter in exactly the same manner as plain cutters. A cup wheel is usually employed in grinding the sides of the cutter teeth. The cutter is

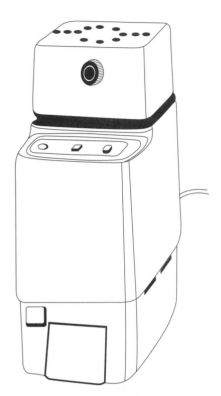

Figure 8-14 Drill bit sharpener.

Figure 8-15 Long-ream grinding attachment for the cutter and tool grinder. This attachment is useful in grinding long lining reamers, extension taps, stay-bolt taps, taper reamers, and boring bars. When concentricity is important, special cutters and gages may be ground without removing them from their arbors. *(Courtesy Cincinnati Milacron Company.)*

Figure 8-16 Clearance-setting dial on the left-hand tailstock of a cutter and tool grinder. Arbor-mounted cutters are quickly adjusted to the desired clearance angle by means of this feature. Clearance-setting dials enable the operator to set predetermined clearance angles conveniently and accurately, regardless of the cutter diameter or type of wheel used. *(Courtesy Cincinnati Milacron Company.)*

Figure 8-17 Radius-grinding attachment. Medium- to large-sized milling cutters requiring an accurate 90° radius on the corner of the teeth can be ground quickly. The attachment consists of four principal elements: a base, swivel plate, adjustable table, and workhead support. A micrometer is included to determine the starting position accurately. The capacity of the attachment is 0 to 1-inch radii, 0 to 12-inch maximum cutter diameter, with 3-inch maximum width of face. *(Courtesy Cincinnati Milacron Company.)*

Figure 8-18 Ample vertical range for going upward to accommodate the exceptionally large diameter face mills, or for going downward to handle complex grinding jobs, is provided by the adjustable wheelhead pile on the cutter and tool grinder. *(Courtesy Cincinnati Milacron Company.)*

Figure 8-19 Face-mill grinding attachment for sharpening face mills up to a diameter of 18 inches more quickly, more easily, and with a higher degree of accuracy. A knurled handwheel at the end of the spindle aids in indexing. The base and the swivel block are both graduated in degrees, permitting the cutter to be swiveled to the desired clearance angle. *(Courtesy Cincinnati Milacron Company.)*

mounted on a stud arbor clamped in the universal workhead, which is swiveled to the required relief angle.

Formed Cutters

A dish-shaped grinding wheel is usually used to grind formed cutters. Either a master form or an index center should be used as a guide for the tooth rest. If these are not available, the tooth rest may be adjusted against the back of the tooth to be ground, prior to which the grinding-wheel face and the center of the cutter have been brought into the same vertical plane. Some form cutters are made with a forward rake, or undercut, tooth. In sharpening these cutters, care must be taken to offset the wheel face, so as to maintain the original rake angle.

Summary

A variety of metal cutting applications can be performed to close tolerances with modern precision tools. Unless the machine has been provided with suitably designed and prepared cutting tools, close tolerances are impossible. The machine, no matter how precise, can be no better than the tool with which it is equipped.

The days of sharpening a twist drill by hand have gone with the appearance of modern tool grinders in up-to-date machine shops. Machine grinding is the more accurate method of sharpening twist drills. A twist drill usually cuts faster, lasts longer, and produces a more accurate hole when ground on a machine than when ground by hand.

The cutting tips of the twist drill should form equal 59° angles with the axis of the drill. Drill manufacturers have found 59° to be the best angle for all-around work. If one lip of a drill is ground at 60°, and the other lip at 59°, it is easy to see that the 60° lip will do all of the cutting.

Machines designed to grind cutters may vary from single types for grinding some of the multitooth cutters to the universal cutters and tool grinding machines that have a range for practically all tool room requirements. Many of the grinding machines have special attachments for grinding various milling machine cutters.

Review Questions

1. What is the recommended cutting angle of a twist drill?
2. What happens when a twist drill is not sharpened at the proper angle, or when one side is at a different angle from the other?

3. What is meant by the lip clearance angle or lip relief angle?

4. If a cutter is used after it becomes _____, it deteriorates rapidly.

5. The cutting lip of the twist drill should form equal angles with the _____ of the drill.

6. The commercial standard angle for a drill bit is _____ degrees.

7. If the lips of the twist drill are ground at 60 degrees and 59 degrees, which one will do all of the work?

8. What is the clearance lip?

9. What is the lip relief angle?

10. Cutter teeth of the side-mill type are ground on the _____ diameter in exactly the same manner as plain cutters.

11. Where is a radius grinding attachment used?

12. A dish-shaped grinding wheel is usually used to grind _____ cutters.

13. What type of grinding wheel is used to thin the web of a drill bit?

Chapter 9

Drills

A *twist drill* is a pointed cutting tool, usually round. It is used for cutting holes in metal or other hard substances and is driven by a machine. As with many other machinist's tools, there is a great variety of twist drills designed to meet all kinds of service.

Drill Standards

Drill diameters have become more and more standardized over the years. Standard drill diameters are classified in a decimal (inch) series that gives the machinist a wide range of twist drills from which to choose. The size of the twist drill is usually stamped on the shank. See Figure 9-1 for parts of the drill bit. Very small drills are not identified by size, but most can be measured with either a micrometer (Figure 9-2) or a drill gage (Figure 9-3). These drills should be kept either in sets or in holders (Figure 9-4 and Figure 9-5). Drills are identified as to size in three ways as follows:

- *Number sizes* from No. 80 (0.0135 inch) to No. 1 (0.228 inch)
- *Letter sizes* from A (0.234 inch) to Z (0.413 inch)
- *Fractional sizes* from ¹⁄₆₄ (0.0156 inch) upward by sixty-fourths of an inch

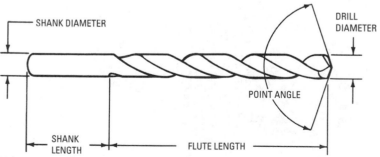

Figure 9-1 Major parts of a twist drill. *(Courtesy Stanley Tools Company.)*

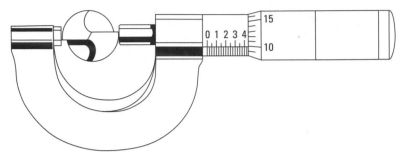

Figure 9-2 Checking a drill bit size using a micrometer.

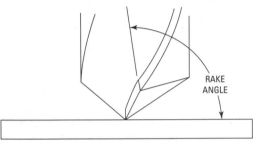

RAKE
ANGLE

Figure 9-3 Rake angle of a drill bit.

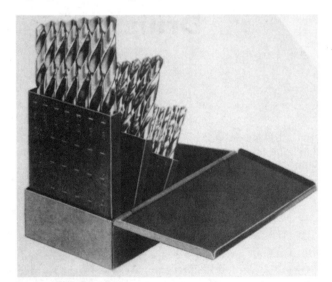

Figure 9-4 Straight-shank drill set. This is a fractional-size set (sizes ¹/₁₆ inch through ¹/₂ inch by sixty-fourths). *(Courtesy National Twist Drill & Tool Company.)*

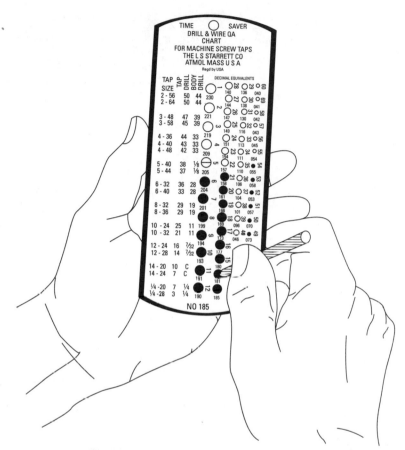

Figure 9-5 Checking drill bit size using a drill and wire gage.

(Courtesy L.S. Starrett Company.)

The fractional sizes that come between the number and letter sizes are shown in Table 9-1. The decimal (inch) diameters of the standard twist drills in number, fraction, and letter sizes are shown in the table.

Twist Drill Terminology

The parts of a twist drill can be seen in Figure 9-1. Note the shank diameter, shank length, flute length, point angle, and drill diameter. By checking these points on the drill bit, you should be able to make sense of any table giving bit diameters and special characteristics.

Table 9-1 Number, Fraction, and Letter Drill Sizes

Drill	Diameter (Inches)	Drill	Diameter (Inches)
80	0.0135	42	0.0935
79	0.0145	3/32	0.0937
1/64	0.0156	41	0.0960
78	0.0160	40	0.0980
77	0.0180	39	0.0995
76	0.0200	38	0.1015
75	0.0210	37	0.1040
74	0.0225	36	0.1065
73	0.0240	7/64	0.1094
72	0.0250	35	0.1100
71	0.0260	34	0.1110
70	0.0280	33	0.1130
69	0.0292	32	0.1160
68	0.0310	31	0.1200
1/32	0.0312	1/8	0.1250
67	0.0320	30	0.1285
66	0.0330	29	0.1360
65	0.0350	28	0.1405
64	0.0360	9/64	0.1406
63	0.0370	27	0.1440
62	0.0380	26	0.1470
61	0.0390	25	0.1495
60	0.0400	24	0.1520
59	0.0410	23	0.1540
58	0.0420	5/32	0.1562
57	0.0430	17/64	0.2656
56	0.0465	H	0.2660
3/64	0.0469	I	0.2720
55	0.0520	J	0.2770
54	0.0550	K	0.2810
53	0.0595	22	0.1570
1/16	0.0625	21	0.1590
52	0.0635	20	0.1610
52	0.0670	19	0.1660
50	0.0700	18	0.1695

Table 9-1 (continued)

Drill	Diameter (Inches)	Drill	Diameter (Inches)
49	0.0730	13	0.1850
48	0.0760	$3/16$	0.1875
$5/64$	0.0781	12	0.1890
47	0.0785	11	0.1910
46	0.0810	10	0.1935
45	0.0820	9	0.1960
44	0.0860	8	0.1990
43	0.0890	7	0.2010
$13/64$	0.2031	N	0.3020
6	0.2040	$5/16$	0.3125
5	0.2055	O	0.3160
4	0.2090	P	0.3230
3	0.2130	$21/64$	0.3281
$7/32$	0.2187	Q	0.3320
2	0.2210	R	0.3390
1	0.2280	$11/32$	0.3437
A	0.2340	S	0.3480
$15/64$	0.2344	T	0.3580
B	0.2380	$23/64$	0.3594
C	0.2420	U	0.3680
D	0.2460	$3/8$	0.3750
E	0.2500	V	0.3770
$1/4$	0.2500	W	0.3860
F	0.2570	$25/64$	0.3906
G	0.2610	X	0.3970
$17/64$	0.2656	Y	0.4040
H	0.2660	$13/32$	0.4062
I	0.2720	Z	0.4130
J	0.2770	$27/64$	0.4219
K	0.2810	$7/16$	0.4375
$9/32$	0.2812	$29/64$	0.4531
L	0.2900	$15/32$	0.4687
M	0.2950	$31/64$	0.4844
$19/64$	0.2969	$1/2$	0.5000

Some drill bits are designed for specific applications. Today the general-purpose drill bit is more or less obsolete. Take a look at the following terminology to make sure you understand the terms used today to refer to specific job-type bits:

- *Screw machine length ("stubby")*—This is a name applied to the shortest drills commonly available. They are designed to be used in automatic drilling machines primarily, and are short for maximum rigidity.

- *Jobbers length*—A term that applies to the length of any standard drill bit up to approximately $^{11}/_{16}$ inch in diameter. It is the drill bit length most often called for by "do-it-yourselfers" and professionals.

- *Extended length* (aircraft extension)—As the name implies, these drill bits are longer than "jobbers" or "taper" length bits. They are used for drilling in hard-to-reach places, such as between studs, or for drilling through extra-thick material.

- *Silver and deming*—This is a name that applies only to $^{1}/_{2}$-inch shank drill bits and refers to a time when drill bits were made specifically for unique machining applications. These drills feature a general-purpose flute design and are recommended for use in portable electrical drills and drill presses.

- *Double-ended drills*—These are drills on which both ends are fluted for a short distance and have a solid center shank for secure chucking. These drills are designed to be reversed quickly in machine tools to reduce changeover time. They often feature a split-point design for fast starting in sheet metal.

The lengths of the drill bits, therefore, increase in size from screw-machine length, or stubby, to jobbers length, to taper length, to extended length, as explained here:

- *Fractional sizes*—These are drills with diameters defined in fractions of an inch. Most drill sets start at $^{1}/_{16}$ inch and go up to 1 inch in diameter, in increments of $^{1}/_{16}$ inch. Fractional sizes are the most commonly specified.

- *Wire gage bits (number sizes)*—Drill numbers correspond approximately to the Stubbs Steel Wire Gage size, an English and American standard. Number series drill bits start at No. 100, the smallest size, and increase in size up to No. 1. Number series drill bits fill the gaps between fractional sizes.

- *Letter sizes*—Letter drill sizes start where the numbers leave off and are also used to fill in the gaps between the fractional sizes. Letters run in size from larger than $\frac{7}{32}$ (A size) up to $\frac{13}{32}$ (Z size). For example, the letters B, C, D, and E are between fractional sizes $\frac{15}{64}$ and $\frac{1}{4}$ inch. Letter size drills fill the need for unique hole sizes.

- *Metric sizes*—As the name implies, metric sizes are a series of drills in diameters corresponding to commonly used metric bolt sizes. Metric drills are sized in millimeters and tenths of a millimeter. Metric drills will become more popular as metric fasteners are more widely used.

Types of Drills

Drills are now manufactured in a great variety of sizes and shapes for a variety of materials and special purposes. In the early days of machine work, the flat drill was used exclusively, but it has given way to the twist drill, which is much more efficient.

Twist drills are made by forging to the approximate size, and milling and grinding the forging to the finished size (Figure 9-6). *Straight-fluted drills* are sometimes used on soft metals, but most drills have spiral flutes. A twist drill is defined as a drill grooved helically along its length for the purpose of clearing itself from the waste material; the borings pass up the grooves as the drill is fed into the work. The parts of the twist drill are the *shank*, the *body*, and the *point* (Figure 9-7).

Shank

The shank of the drill fits into the chuck of the machine that revolves the drill. Drills with straight shanks (Figure 9-8) are held in a drill chuck, which has three jaws that grip the drill.

Taper-shank drills mount directly into the taper holder of the drill press spindle, or in a sleeve. The tang of the taper-shank drill fits into a slot in the spindle to prevent the drill from slipping (Figure 9-7). The automotive series of drills has tangs for use with split-sleeve drill drivers. (Figure 9-9).

Body

The body of a twist drill extends from the shank to the point. The *flutes* are helical grooves running along opposite sides of the drill. A straight-fluted drill may be used for free machining brass, bronze, or other soft materials (Figure 9-10). *Three-* or *four-fluted drills* are especially adapted for enlarging punched, cored, or drilled holes (Figure 9-11 and Figure 9-12). The *margin* of a drill lies along the

Figure 9-6 Milling the spiral flutes on a drill.

(Courtesy National Cincinnati Milacron Company.)

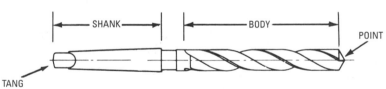

Figure 9-7 A general-purpose taper-shank twist drill.

(Courtesy National Twist Drill & Tool Company.)

Figure 9-8 A general-purpose straight-shank twist drill.

(Courtesy National Twist Drill & Tool Company.)

Figure 9-9 Taper-length, tanged, automotive series, straight-shank twist drill regularly furnished with tangs for use with split-sleeve drill drivers. *(Courtesy National Twist Drill & Tool Company.)*

Figure 9-10 Straight-fluted drill for free machining brass, bronze, or other soft materials, particularly on screw machines. Also suitable for drilling thin sheet material because of lack of tendency to "hog." *(Courtesy Morse Twist Drill & Machine Company.)*

Figure 9-11 Straight-shank, three-fluted core drill. *(Courtesy National Twist Drill & Tool Company.)*

Figure 9-12 Straight-shank, four-fluted core drill. *(Courtesy National Twist Drill & Tool Company.)*

entire length of the flute. This determines the correct size of the hole. A measurement with a micrometer across both margins gives the size of the drill.

The *body clearance* is immediately in back of the margin. This reduced diameter decreases the friction between the drill and the wall of the hole so that the drill does not bind.

The *land* is the portion of the drill body that is not cut away by the flutes. This includes both the body clearance and the margin.

Back taper allows a slight clearance for the drill in the hole. This is a slight taper in the body of the drill, from the cutting end to the tang end, made by constructing the body of the drill slightly smaller near the shank end.

In *low-helix drills* (Figure 9-13), the flutes are made wider than those of regular drills to prevent the chips from catching and clogging, or to increase the chip space. The flute core diameters are less on low-helix drills. They require less pressure, give easier penetration, and develop less heat than regular drills. *High-helix drills* have higher cutting rake and improved chip-conveying properties for use in aluminum and some plastics (Figure 9-14).

Figure 9-13 Low-helix drill used extensively in screw machines making parts from screw stock and brass. *(Courtesy National Twist Drill & Tool Company.)*

Figure 9-14 High-helix drill designed with higher cutting rake and improved chip-conveying properties for use on such materials as aluminum, diecasting alloys, and some plastics.
(Courtesy National Twist Drill & Tool Company.)

The length of the body of twist drills may vary. *Screw-machine length twist drills* have a shorter body (Figure 9-15). *Center drills* (Figure 9-16) and *starting drills* (Figure 9-17) are also used in screw-machine operations. *Left-hand drills* are used where the spindle of the machine rotates in a left-hand direction.

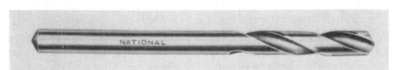

Figure 9-15 Screw-machine length straight-shank twist drill.
(Courtesy National Twist Drill & Tool Company.)

Figure 9-16 Center drill.
(Courtesy National Twist Drill & Tool Company.)

Figure 9-17 Starting drill. *(Courtesy National Twist Drill & Tool Company.)*

Multidiameter drills of the step drill type have two or more diameters produced by successive steps on the lands of the drill. The steps are separated by square or angular shoulders. In the *oil-hole drill* (Figure 9-18), there is a hole through the solid metal for conveying the lubricant to the point. When drilling in cast iron, air is sometimes used to blow out the chips and to keep the drill cool.

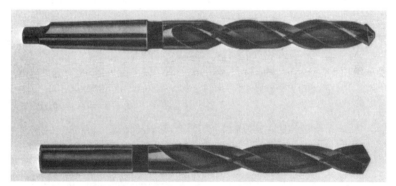

Figure 9-18 Oil-hole twist drills for production work in all types of materials on screw machines or turret lathes.
(Courtesy National Twist Drill & Tool Company.)

Point

The point is the cone-shaped end or cutting part of the twist drill. The extreme tip of the drill, which forms one sharp edge, is the *dead center*, or web (Figure 9-19). The dead center, or web, acts as a flat drill, and cuts its own hole in the work. This is the reason it is common practice to drill a lead hole in the work first. The lead hole

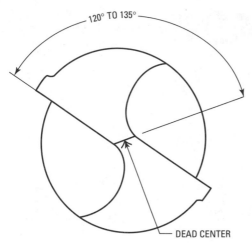

120° TO 135°

DEAD CENTER

Figure 9-19   Note the location of the dead center on the end of the drill bit.

provides clearance for the dead center of a twist drill. Thus, the larger drills are kept from running off center, and less feeding pressure is required.

The cutting edges extending from the dead center to the periphery of the drill point are called *lips*. The two cutting edges, or lips, make a standard (helix) angle of 59° with the axis of the drill body. Thus, the included angle of the drill point is 118°. This angle may be varied with the type of material being drilled (Figure 9-20).

Figure 9-20   Drill used principally for drilling molded plastics. Also used for drilling hard rubber, wood, and aluminum and magnesium alloys. *(Courtesy National Twist Drill & Tool Company.)*

Twist drills for general-purpose work usually have a *lip clearance* of 12° to 15° at the extreme diameter of the drill. Without a lip clearance, the twist drill could not cut because the metal immediately behind the lip would rub on the bottom of the newly drilled hole.

Drill-Bit Point Design

The point of a drill bit indicates its application. For example, a common drill bit has a point angle of 118° (see Figure 9-21). Figure 9-21A shows how the flutes have a moderate spiral going up the shank, which indicates that the drill bit is used for general drilling purposes of shallow depth in soft metals, wood, and so on. If the point angle is flatter (such as 135°, as shown in Figure 9-21B), the drill is intended to be used in tough metals where a smaller chip is taken off with each rotation. The web, the center portion of the drill bit, is very thick (see Figure 9-21C). If a drill bit is to be used in sheet metal, where a lot of force pushing the drill into the metal can cause denting, the point of the bit is sometimes ground in the web area. Grinding the web reduces the width of the chisel point (see Figure 9-21D). This minimizes the amount of force needed to push the drill bit into the metal. The result is a split-point design that reduces the chance of denting the metal surfaces and minimizes drill walking.

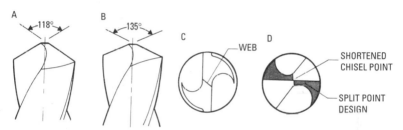

Figure 9-21 Web and split-point twist drills. *(Courtesy Stanley Tools Company.)*

Special-Purpose Drills

Manufacturers have developed twist drills for special purposes. The shank, body, and point have been altered to meet the various needs.

Aircraft drills (Figure 9-22) are designed for drilling light sections of high-strength, high-temperature alloys and similar materials used in aircraft and missile construction. Maximum tool rigidity is built in by using heavy-duty construction and short flute lengths on regular length drills.

Die drills may be used for drilling hard steel. The drills shown in Figure 9-23 are carbide drills. Other twist drills for special purposes are the *metalworking drill* (Figure 9-24), which has a ¼-inch–diameter shank adapted to fit the chucks of electric and hand drills, and the *Silver and Deming drill* (Figure 9-25), with a ½-inch–diameter shank. The *bonding drill* (Figure 9-26) is used for drilling holes for bonding wire in track and signal work.

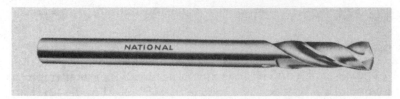

Figure 9-22 Aircraft drill used in aircraft and missile construction. These drills are effective in drilling in the higher ranges of work material hardness. Point construction produces the highest possible penetration and thrust reduction with maximum cutting lip support. *(Courtesy National Twist Drill & Tool Company.)*

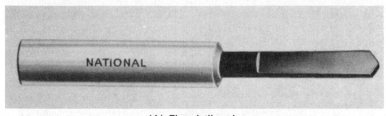

(A) Flat-drill style

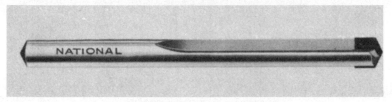

(B) Carbide-tipped style

Figure 9-23 Die drills used for drilling hard steel. *(Courtesy National Twist Drill & Tool Company.)*

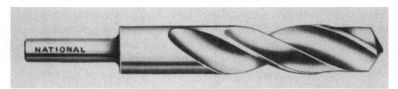

Figure 9-24 Metalworking drill with ¼-inch–diameter shank. Fits electric and hand drills with ¼-inch chucks. Special notched point permits easy drilling in all metals. *(Courtesy National Twist Drill & Tool Company.)*

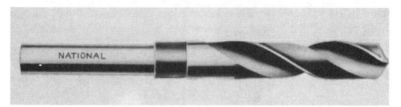

Figure 9-25 Silver and Deming drill with ½-inch–diameter shank.
(Courtesy National Twist Drill & Tool Company.)

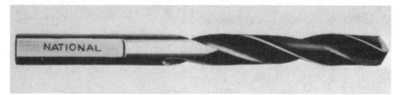

Figure 9-26 Bonding drill for drilling holes for bonding wire in track and signal work. (Courtesy National Twist Drill & Tool Company.)

The *combined drill and countersink* is designed for properly finishing the center holes of work to be turned on a lathe (Figure 9-27 and Figure 9-28). The angle of the tapered part of the drill is 60° to conform to the taper of the lathe centers. The small extension bores out a small hole that extends beyond the point of the lathe center, protecting it from injury.

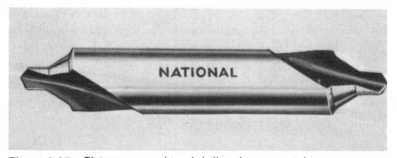

Figure 9-27 Plain-type combined drill and countersink. (Courtesy National Twist Drill & Tool Company.)

Socket and Sleeve

Most drill presses are equipped with a spindle having a Morse taper. The chuck for the drill press has a taper shank and tang that

Figure 9-28 Bell-type combined drill and countersink. *(Courtesy National Twist Drill & Tool Company.)*

fit into the taper in the drill press spindle. The chuck shank has a standard Morse taper. Of course, the chuck may be used to hold the straight-shank twist drills up to ½ inch in size.

Taper-shank twist drills may be inserted directly into drill press spindles, or they may be inserted in various holding devices having a taper hole. The tang of the twist drill fits into a small slotted recess inside the drill press spindle, above the tapered hole, and keeps the drill from slipping. If the taper hole in the spindle is too large for the shank of the twist drill or chuck, the taper-shank tool is held in a steel sleeve (Figure 9-29). The sleeve has a Morse taper on the outside so that it will fit into the tapered hole in the spindle. The tang on the end of the sleeve fits in the slot in the spindle. The taper on the drill shank fits in the tapered hole on the inside of the sleeve.

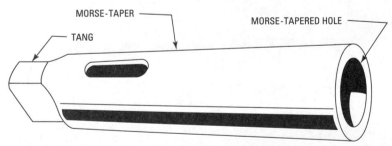

Figure 9-29 Steel sleeve for taper-shank twist drill.

The steel socket (Figure 9-30) is different from the sleeve in that a larger taper-shank drill may be inserted into the socket than may be inserted into the spindle or sleeve.

Figure 9-30   Steel socket for taper-shank twist drill. *(Courtesy National Twist Drill & Tool Company.)*

Using the Twist Drill

The versatility of the drill bit makes it one of the most widely used power tool accessories in use today.

Twist drills will stand more strain in proportion to their size than any other small cutting tool. When a drill bit chips, breaks, burns, or does not cut properly, it is probably because it is being abused. A few suggestions can be utilized to eliminate many drilling problems and ensure maximum drill life.

Secure the Work

To avoid injury or damage, the work should be securely anchored. When drilling, the bit should be perpendicular to the work. Wearing safety goggles minimizes the chance of eye injury.

Cutting Oil

The use of fluids when drilling metals increases the life of the drill bit. Fluids that aid in drilling are as follows:

- *Oil*—Used for steel
- *Soluble oil*—Used for bronze, soft steel, wrought iron
- *Kerosene*—Used for aluminum, aluminum alloys

Speeds and Feeds

Speed refers to the revolutions per minute (rpm) of the drill press spindle. The speed of a twist drill is calculated at the circumference of the drill, or the *peripheral speed*. Peripheral speed, then, is the speed of travel of a point on the largest diameter in surface feet per minute (sfpm)—not in revolutions per minute.

Extensive experimentation has determined a recommended cutting speed (sfpm) for each of the various materials as shown in Table 9-2.

Table 9-2 Recommended Drilling Speeds for Materials with High-Speed Drills

Material	Recommended Speed in Surface Feet per Minute (sfpm)
Aluminum and alloys	200–300
Bakelite	100–150
Plastics	100–150
Brass and bronze, soft	200–300
Bronze, high tensile	70–100
Cast iron, chilled	30–40
Cast iron, hard	70–100
Cast iron, soft	100–150
Magnesium and alloys	250–400
Malleable iron	80–90
Monel, metal	40–50
Nickel	40–60
Steel, annealed (.4 to .5 percent C)	60–70
Steel, forgings	50–60
Steel, machine (.2 to .3 percent C)	80–110
Steel, manganese (15 percent Mn)	15–25
Steel, soft	80–100
Steel, stainless (free machining)	60–70
Steel, stainless (hard)	30–40
Steel, tool (1.2 percent C)	50–60
Slate, marble, and stone	15–25
Wrought iron	50–60
Wood	300–400

After the operator has determined the recommended cutting speed for the material to be worked, he or she must convert the surface feet per minute (sfpm) to revolutions per minute (rpm) at which the drill press spindle must turn for the size of drill being used. The operator may either make the calculations or consult a table.

If you prefer to make the calculations, use the following formula:

$$V = \frac{\pi DN}{12}$$

where the following is true:

V = velocity or cutting speed in feet per minute (sfpm)

D = diameter of twist drill

N = revolutions per minute (rpm)

π = 3.1416

For example, from Table 9-2, the cutting speed for cast iron is 100 sfpm, and the operator wanted to use a ½-inch twist drill. Substituting in the formula

$$100 = \frac{(3.1416)(.5)(N)}{12}$$

$$1200 = 1.5708\,N$$

$$\frac{1200}{1.5708} = N$$

$764 = N$, or revolutions per minute (r/min)

the calculated desired speed, 764 rpm, is the same as the speed given in Table 9-3.

Table 9-3   Conversion Table for Surface Feet Per Minute (sfpm) to Revolutions Per Minute (rpm)

Size				Surface Feet per Minute (sfpm)						
Fraction	Wire Gage	Machine Screw Tap	Decimal Equivalent	20	25	30	35	40	45	50
	80		0.0135	5659	7073	8488	9858	11,317	12,732	14,146
	79		0.0145	5269	6586	7903	9179	10,538	11,855	13,172
¹⁄₆₄			0.0156	4897	6121	7345	8531	9794	11,018	12,242
	78		0.016	4775	5968	7162	8319	9549	10,743	11,937
	77		0.018	4244	5306	6367	7394	8489	9550	10,611
	76		0.020	3820	4775	5730	6655	7639	8594	9549
	75		0.021	3638	4547	5457	6338	7276	8185	9095
	74		0.0225	3395	4244	5092	5915	6790	7639	8487
	73		0.024	3183	3979	4775	5546	6367	7163	7958

(continued)

Table 9-3   (continued)

Size				Surface Feet per Minute (sfpm)						
Fraction	Wire Gage	Machine Screw Tap	Decimal Equivalent	20	25	30	35	40	45	50
	72		0.025	3056	3820	4584	5324	6112	6875	7639
	71		0.026	2938	3673	4407	5119	5876	6612	7345
	70		0.028	2728	3410	4092	4753	5456	6138	6820
	69		0.0293	2607	3259	3911	4542	5215	5867	6518
	68		0.031	2464	3081	3697	4293	4929	5545	6161
$1/32$			0.0312	2448	3061	3673	4266	4897	5509	6121
	67		0.032	2387	2984	3581	4159	4775	5371	5968
	66		0.033	2315	2893	3472	4033	4629	5208	5787
	65		0.035	2183	2728	3274	3802	4365	4911	5456
	64		0.036	2122	2653	3183	3697	4244	4775	5306
	63		0.037	2065	2581	3097	3597	4130	4646	5162
	62		0.038	2011	2513	3016	3503	4021	4524	5027
	61		0.039	1959	2448	2938	3412	3917	4407	4897
	60		0.040	1910	2387	2865	3327	3820	4297	4775
	59		0.041	1863	2329	2795	3246	3726	4192	4658
	58		0.042	1819	2274	2728	3169	3638	4093	4547
	57		0.043	1777	2221	2665	3096	3554	3998	4442
	56		0.0465	1643	2054	2465	2863	3286	3697	4108
$3/64$			0.0469	1629	2036	2443	2837	3257	3665	4072
	55		0.052	1469	1836	2204	2559	2938	3305	3673
	54		0.055	1389	1736	2083	2420	2778	3125	3472
	53		0.059	1295	1619	1942	2556	2590	2913	3237
		0	0.060	1273	1592	1910	2219	2547	2865	3184
$1/16$			0.0625	1222	1528	1833	2129	2445	2750	3056
	52		0.0635	1203	1504	1805	2096	2406	2707	3008
	51		0.067	1141	1426	1711	1987	2281	2566	2581
	50		0.070	1092	1365	1638	1902	2183	2456	2729
	49	1	0.073	1047	1308	1570	1823	2093	2355	2616
	48		0.076	1005	1257	1508	1751	2011	2262	2513
$5/64$			0.0781	978	1222	1467	1704	1956	2200	2445
	47		0.0785	973	1217	1460	1696	1947	2190	2433

Table 9-3   (continued)

Size				Surface Feet per Minute (sfpm)						
Fraction	Wire Gage	Machine Screw Tap	Decimal Equivalent	20	25	30	35	40	45	50
	46		0.081	943	1179	1415	1644	1887	2123	2359
	45		0.082	932	1165	1398	1624	1864	2097	2330
	44	2	0.086	888	1111	1333	1548	1777	1999	2221
	43		0.089	859	1073	1288	1496	1717	1932	2147
	42		0.0935	817	1022	1226	1424	1635	1839	2044
³/₃₂			0.0938	814	1018	1222	1419	1629	1832	2036
	41		0.096	796	995	1194	1387	1592	1791	1990
	40		0.098	779	974	1169	1358	1558	1753	1948
		3	0.099	772	964	1157	1344	1543	1736	1929
	39		0.0995	768	960	1152	1338	1536	1727	1919
	38		0.1015	752	941	1129	1311	1505	1693	1881
	37		0.104	735	919	1102	1280	1470	1654	1837
	36		0.1065	717	897	1076	1250	1435	1614	1793
⁷/₆₄			0.1094	698	873	1047	1216	1396	1571	1746
	35		0.110	694	868	1042	1210	1389	1562	1736
	34		0.111	688	860	1032	1199	1377	1549	1721
	33		0.113	676	845	1014	1178	1352	1521	1690
		4	0.115	664	831	997	1158	1329	1495	1662
	32		0.116	659	823	988	1147	1317	1482	1646
	31		0.120	636	795	955	1109	1273	1432	1591
⅛		5	0.125	611	764	917	1065	1222	1375	1528
	30		0.1285	594	743	892	1035	1189	1337	1486
	29		0.136	561	702	842	978	1123	1263	1404
		6	0.138	554	692	831	965	1108	1246	1385
	28		0.1405	544	680	816	948	1088	1224	1360
⁹/₆₄			0.1406	543	679	815	946	1086	1222	1358
	27		0.144	530	663	795	924	1060	1193	1325
	26		0.147	519	649	779	905	1039	1169	1299
	25		0.1495	511	639	767	890	1022	1150	1278
	24		0.152	503	628	754	876	1005	1131	1257
	23		0.154	496	620	744	864	992	1116	1239

(continued)

Table 9-3 *(continued)*

Fraction	Wire Gage	Machine Screw Tap	Decimal Equivalent	20	25	30	35	40	45	50
Size				Surface Feet per Minute (sfpm)						
5/32			0.1562	489	611	733	852	978	1100	1222
	22		0.157	487	608	730	848	973	1095	1217
	21		0.159	481	601	721	837	961	1081	1201
	20		0.161	474	593	712	826	949	1067	1186
		8	0.164	466	583	699	812	932	1049	1165
	19		0.166	460	575	690	801	920	1035	1150
	18		0.1695	451	563	676	785	901	1014	1127
11/64			0.172	444	555	666	773	888	999	1110
	17		0.173	442	552	662	769	883	994	1104
	16		0.177	432	540	647	752	863	971	1079
	15		0.180	425	531	637	740	850	956	1062
	14		0.182	419	524	629	731	839	944	1049
	13		0.185	413	517	620	720	827	930	1033
3/16			0.1875	407	509	611	709	814	916	1018
	12		0.189	404	505	606	704	808	909	1010
		10	0.190	402	502	603	700	804	904	1005
	11		0.191	400	500	600	697	801	901	1001
	10		0.1935	395	494	592	688	790	889	987
	9		0.196	390	487	584	679	779	877	974
	8		0.199	384	480	576	669	769	865	961
	7		0.201	380	476	571	663	761	856	951
13/64			0.2031	376	470	564	655	752	846	940
	6		0.204	374	468	561	652	749	842	936
	5		0.2055	372	465	558	648	744	837	930
	4		0.209	365	456	548	636	730	822	913
	3		0.213	358	448	537	624	717	806	896
		12	0.216	354	442	531	616	707	796	884
7/32			0.2188	349	436	524	608	698	786	873
	2		0.221	345	432	518	602	691	777	863
	1		0.228	335	419	503	584	671	755	838
			0.234	326	408	489	568	652	734	816

Table 9-3 (continued)

Size				Surface Feet per Minute (sfpm)						
Fraction	Wire Gage	Machine Screw Tap	Decimal Equivalent	20	25	30	35	40	45	50
¹⁵⁄₆₄			0.2344	326	408	489	568	652	734	816
			0.238	321	401	481	559	642	722	802
		14	0.242	316	394	473	550	631	710	789
	D		0.246	311	389	466	542	622	700	777
¼	E		0.250	306	382	458	532	611	688	764
	F		0.257	297	371	446	518	594	669	743
	G		0.261	293	366	439	510	585	658	731
¹⁷⁄₆₄			0.2656	288	360	432	502	576	648	720
	H		0.266	287	359	431	500	574	646	718
	I		0.272	281	351	422	490	562	633	703
	J		0.277	276	345	414	480	552	621	689
	K		0.281	272	340	408	474	544	612	680
⁹⁄₃₂			0.2815	271	339	407	472	542	610	678
	L		0.290	264	329	395	459	527	593	659
	M		0.295	259	324	388	451	518	583	647
¹⁹⁄₆₄			0.2969	257	322	386	449	515	579	644
	N		0.302	253	316	379	441	506	569	632
⁵⁄₁₆			0.3125	244	306	367	426	489	550	611
	O		0.316	241	302	362	421	483	543	604
	P		0.323	237	296	355	413	474	533	592
²¹⁄₆₄			0.3281	233	291	350	406	466	524	583
	Q		0.332	230	287	345	401	460	517	575
	R		0.339	225	282	338	393	451	507	563
¹¹⁄₃₂			0.3438	222	278	333	387	445	500	556
	S		0.348	219	274	329	382	439	493	548
	T		0.358	213	266	320	371	426	480	533
²³⁄₆₄			0.3594	212	265	319	370	425	478	531
	U		0.368	208	260	312	362	416	468	519
³⁄₈			0.375	204	255	306	355	408	459	510
	V		0.377	202	253	304	353	405	456	506
	W		0.386	198	247	297	345	396	445	495

(continued)

Table 9-3 (continued)

Size				Surface Feet per Minute (sfpm)						
Fraction	Wire Gage	Machine Screw Tap	Decimal Equivalent	20	25	30	35	40	45	50
25/64			0.3906	196	244	293	341	391	440	489
	X		0.397	193	241	289	335	385	433	481
	Y		0.404	189	237	284	330	379	426	474
13/32			0.4062	188	235	282	327	376	423	470
	Z		0.413	185	231	277	322	370	416	462
7/16			0.4375	175	219	262	305	350	394	437
15/32			0.4688	163	203	244	283	325	366	407
1/2			0.500	153	191	229	266	306	344	382
9/16			0.5625	136	170	204	237	272	306	340
5/8			0.625	122	153	183	213	244	275	306
11/16			0.6875	111	138	166	193	221	249	277
3/4			0.7500	105	127	152	177	203	229	254
13/16			0.8125	94	117	141	164	188	211	235
7/8			0.875	87	109	131	152	174	196	218
15/16			0.9375	82	102	123	142	163	184	204
1			1.000	76	95	115	133	153	172	191
1 1/8			1.125	68	85	102	118	136	153	170
1 1/4			1.250	61	76	92	106	122	138	153
1 3/8			1.375	56	70	84	97	112	125	139
1 1/2			1.500	51	64	77	89	102	115	128
1 5/8			1.625	47	59	71	83	95	107	118
1 3/4			1.750	44	54	65	76	87	98	109
1 7/8			1.875	40	51	61	71	81	91	101
2			2.000	38	48	57	67	76	86	95
2 1/2			2.500	31	38	46	53	61	69	76
3			3.000	25	32	38	44	50	57	63
3 1/2			3.500	22	28	33	39	44	50	55
4			4.000	19	24	29	33	38	43	48
4 1/2			4.500	17	21	25	29	34	38	42
5			5.000	15	19	23	27	31	34	38
5 1/2			5.500	14	17	21	24	28	31	34
6			6.000	13	16	19	23	26	29	32
6 1/2			6.500	11	14	17	20	23	26	29

Table 9-3   (continued)

Size				Surface Feet per Minute (sfpm)						
Fraction	Wire Gage	Machine Screw Tap	Decimal Equivalent	20	25	30	35	40	45	50
7			7.000	12	13	16	19	21	24	27
8			8.000	10	12	15	17	20	22	25
9			9.000	08	11	13	15	17	19	21
10			10.000	08	10	11	13	15	17	19
11			11.000	07	09	10	12	14	15	17
12			12.000	06	08	09	11	12	14	15

A simple formula for calculating the proper speed to run a twist drill for a particular metal is as follows:

$$r/min = \frac{4S}{D}$$

where

S = cutting speed or sfpm of the metal

D = diameter of twist drill

$$r/min = \frac{4(100)}{.5} = \frac{400}{.5} = 800$$

The roughly calculated speed, 800 rpm, compares favorably with the speed calculated by the formula (or obtained from the table), 764 rpm, as a means of obtaining the desired rpm for the drill press spindle. In many instances, time can be saved by performing the rough calculation.

The feed of a twist drill refers to the downward movement into the work during each rpm of the drill press spindle. Generally, the larger the drill, the heavier the feed that may be used. The drill feed per rpm is given in Table 9-4.

The feeds should generally be less than those shown in the table for alloys and hard steels. A heavier feed may usually be used for cast iron, brass, and aluminum. Of course, a proper coolant is necessary to maintain the recommended speeds and feeds.

Table 9-4   Drill Feed in Inches per Revolution

Reference Symbol	Diameter of Drill—Inches			
	Under ⅛	⅛ to ¼	¼ to I	Over I
L—Light	0.001	0.002	0.003	0.006
M—Medium	0.0015	0.003	0.006	0.012
H—Heavy	0.0025	0.005	0.010	0.025

Clearance Drills

A clearance drill is used to drill a hole of sufficient size so that a bolt or screw will pass through it. This drill creates a hole with a clearance for the outside (major) diameter of the bolt or screw. The difference between the clearance drill size and the thread is referred to as the clearance. From Table 9-5, clearance drill equals the diameter of the bolt or screw plus the clearance:

Clearance drill = diameter of thread + clearance

$$CD = 1'' + \frac{1}{64}$$

$$CD = 1\frac{1}{64}''$$

Table 9-5   Clearance Drill Sizes

Thread Sizes		Outside Diameter (inches)	Root Diameter (inches)	Clearance Drill (Inches)		
UNC NC (ISS)	UNF NF (SAE)			Size	Decimal Equivalent	Clearance (Inc.)
	#0–80	0.0600	0.0438	#51	0.0670	0.0070
#1–64		0.0730	0.0527	#47	0.0785	0.0055
	#1–72	0.0730	0.0550	#47	0.0785	0.0055
#2–56		0.0860	0.0628	#42	0.0935	0.0075
	#2–64	0.0860	0.0657	#42	0.0935	0.0075
#3–48		0.0990	0.0719	#36	0.1065	0.0075
	#3–56	0.0990	0.0758	#36	0.1065	0.0075
#4–40		0.1120	0.0795	#31	0.1200	0.0080
	#4–48	0.1120	0.0849	#31	0.1200	0.0080
#5–40		0.1250	0.0925	#29	0.1360	0.0110
	#5–44	0.1250	0.0955	#29	0.1360	0.0110
#6–32		0.1380	0.0974	#25	0.1495	0.0115
	#6–40	0.1380	0.0155	#25	0.1495	0.0115

Table 9-5   (continued)

Thread Sizes		Outside	Root	Clearance Drill (Inches)		
UNC NC (ISS)	UNF NF (SAE)	Diameter (inches)	Diameter (inches)	Size	Decimal Equivalent	Clearance (Inc.)
#8–32		0.1640	0.1234	#16	0.1770	0.0130
	#8–36	0.1640	0.1279	#16	0.1770	0.0130
#10–24		0.1900	0.1359	$^{13}\!/_{64}$	0.2031	0.0131
	#10–32	0.1900	0.1494	$^{13}\!/_{64}$	0.2031	0.0131
#12–24		0.2160	0.1619	$^{7}\!/_{32}$	0.2187	0.0027
	#12–28	0.2160	0.1696	$^{7}\!/_{32}$	0.2187	0.0027
¼"–20		0.2500	0.1850	$^{17}\!/_{64}$	0.2656	0.0156
	¼"–28	0.2500	0.2036	$^{17}\!/_{64}$	0.2656	0.0156
⁵⁄₁₆"–18		0.3125	0.2403	$^{21}\!/_{64}$	0.3281	0.0156
	⁵⁄₁₆"–24	0.3125	0.2584	$^{21}\!/_{64}$	0.3281	0.0156
⅜"–16		0.3750	0.2938	$^{25}\!/_{64}$	0.3906	0.0156
	⅜"–24	0.3750	0.3209	$^{25}\!/_{64}$	0.3906	0.0156
⁷⁄₁₆"–14		0.4375	0.3447	$^{29}\!/_{64}$	0.4531	0.0156
	⁷⁄₁₆"–20	0.4375	0.3725	$^{29}\!/_{64}$	0.4531	0.0156
½"–13		0.5000	0.4001	$^{33}\!/_{64}$	0.5156	0.0156
	½"–20	0.5000	0.4350	$^{33}\!/_{64}$	0.5156	0.0156
⁹⁄₁₆"–12		0.5625	0.4542	$^{37}\!/_{64}$	0.5781	0.0156
	⁹⁄₁₆"–18	0.5625	0.4903	$^{37}\!/_{64}$	0.5781	0.0156
⅝"–11		0.6250	0.5069	$^{41}\!/_{64}$	0.6406	0.0156
	⅝"–18	0.6250	0.5528	$^{41}\!/_{64}$	0.6406	0.0156
¾"–10		0.7500	0.6201	$^{49}\!/_{64}$	0.7656	0.0156
	¾"–10	0.7500	0.6688	$^{49}\!/_{64}$	0.7656	0.0156
⅞"–9		0.8750	0.7307	$^{57}\!/_{64}$	0.8906	0.0156
	⅞"–9	0.8750	0.7822	$^{57}\!/_{64}$	0.8906	0.0156
1"–8		1.0000	0.8376	$1^{1}\!/_{64}$	1.0156	0.0156
	1"–14	1.0000	0.9072	$1^{1}\!/_{64}$	1.0156	0.0156

Table 9-6 shows the drill geometry for high-speed steel twist drills.

TABLE 9-6   Drill Geometry for High-Speed Steel Twist Drills

Material	Point Angle	Lip Relief Angle	Chisel Edge Angle	Helix Angle	Point Grind
Aluminum alloys	90–118	12–15	125–135	24–48	Standard
Magnesium alloys	70–118	12–15	120–135	30–45	Standard
Copper alloys	118	12–15	125–135	10–30	Standard

(continued)

TABLE 9-6   Drill Geometry for High-Speed Steel Twist Drills

Material	Point Angle	Lip Relief Angle	Chisel Edge Angle	Helix Angle	Point Grind
Steels	118	10–15	125–135	24–32	Standard
High-strength steels	118–135	7–10	125–135	24–32	Crankshaft
Stainless steels, low-strength	118	10–12	125–135	24–32	Standard
Stainless steels, high-strength	118–135	7–10	120–130	24–32	Crankshaft
High-temperature alloys	118–135	9–12	125–135	15–30	Crankshaft
Refractory alloys	118	7–10	125–135	24–32	Standard
Titanium alloys	118–135	7–10	125–135		Crankshaft
Cast irons	118	8–12	125–135	24–32	Standard
Plastics	60–90	7	120–135	29	Standard

Drill Problems and Causes

A variety of causes can be pinpointed to drill problems, as shown in Table 9-7.

Table 9-7   Drill Problems and Causes

Problem	Cause
Drill will not cut	Dull drill bit; work piece too hard; drill speed too fast.
Hole rough	Too much pressure on the drill; work not secure; dull or improperly sharpened drill; cutting oil needed.
Hole oversize	Drill bit bending; improper size drill; work not secure; drill not sharpened properly.
Drill bit walks	No center punch; no pilot hole to lead drill into place; improper drill point.
Drill bit grabs work	Improper drill point; work not secure; too much speed or pressure.
Drill bit burning	Too much speed or pressure; flutes clogged; work too hard; cutting oil needed; dull drill bit.
Drill breaks when drilling	Flutes clogged with chips; work not secure; improper drill type; too much pressure or pressure not applied on centerline of bit.

Summary

Drills are manufactured in many sizes and shapes for a variety of materials and special purposes. Twist drills are made by forging to the approximate size, and then milling and grinding to the finished size. Straight-fluted drills are sometimes used on soft metals, but most drills have spiral flutes.

The parts of the twist drill are the shank, the body, and the point. The shank of the drill fits into the chuck of the machine that revolves the drill. This part can be straight or tapered, depending on the type of drill. The body of the drill extends from the shank to the point. The flutes or grooves run along opposite sides of the body. The point is the cone-shaped end or cutting part of the twist drill. The cutting edges extending from dead center to the periphery of the drill point are called lips. The lips are formed at 59°, which is determined as the best cutting angle.

Manufacturers have developed twist drills for special purposes. The shank, body, and point have been altered to meet various needs. For example, aircraft drills are designed for drilling light sections of high-strength, high-temperature alloys and similar materials used in aircraft and missile construction.

Die drills may be used for drilling hard steel. Metalworking drills with a ¼-inch–diameter shank fit electric and hand drills with ¼-inch chucks. A special notched point permits easy drilling in all metals. Some carbide tipped drill bits are made with a smaller shank for drilling in concrete.

Most drill presses are equipped with a spindle having a Morse taper. The chuck for the drill press has a taper shank and tang that fit into the taper in the drill press spindle. The chuck shank has a standard Morse taper. Taper shank twist drills may be inserted directly into drill press spindles, or they may be inserted in various holding devices that have a taper hole. The tang of the twist drill fits into a small slotted recess inside the drill press spindle above the taper hole and keeps the drill from slipping.

Review Questions

1. What are the shank, body, and point of a twist drill?
2. What is an oil-hole drill, and why is it used?
3. Explain the purpose of the multidiameter drill.
4. What are the lips of the drill?
5. How are drill sizes identified?
6. A twist drill is a pointed _____ tool.

7. Standard drill diameters are classified in a _____ series.

8. What size is an "A" drill bit?

9. What is the range in sizes for the letter size drill bits?

10. What is the flute length of a drill bit?

11. What is the rake angle?

12. How do you check to make sure the drill bit is the correct size?

13. Identify the following terms:
    a. Screw machine length
    b. Jobbers length
    c. Extended length
    d. Double ended
    e. Silver and Deming

14. Letter drill sizes start where the _____ leave off and are also used to fill the gaps between fractional sizes.

15. Drill numbers for wire gage bits correspond approximately to the _____ Steel Wire Gage.

16. The _____ of a twist drill extends from the shank to the point.

17. What is the tang of a twist drill bit?

18. Identify the following terms: body clearance, land, and back taper.

19. The _____ of a drill bit indicates its application.

20. What is the purpose of an aircraft drill?

21. What is the purpose of a bonding drill?

22. What is meant by Morse taper?

23. What type of cutting oil is needed for drilling bronze?

24. The feed of the twist drill refers to the _____ movement into the work.

25. What is the cause if you have a rough hole when you use a drill bit?

Chapter 10

Reamers

A *reamer* is a precision tool designed to finish (to a specified diameter) a hole that has been produced either by drilling or by other means. It is quite impossible to drill a hole to an exact standard diameter. Where precision is required, a hole is first drilled to a few thousandths of an inch less than the desired size and then reamed to exact size.

A reamer is both fluted and slightly tapered in its construction. The blades are worked out of the solid metal by planing or milling on a machine. The flutes are then backed off to give a cutting edge. The size of a reamer can be determined with a micrometer, measuring across two opposite cutting edges.

Types of Reamers

There are many types of reamers in use. A reamer has its place in industry as a finishing and sizing tool. The use of reamers has been somewhat reduced by the introduction of the internal grinder, but work that must undergo machining in the lathe, for example, may be finished and brought to size with a reamer. Reamers may be used on work in the drill press without removing the work from the drill press, and on milling machines. Of course, hand reamers are used extensively.

Hand Reamers

Hand reamers have a square end to engage the hand wrench. The fluted or cutting part on hand reamers is slightly tapered on the end to facilitate starting the reamer properly. The shank is 0.005 inch under the size of the reamer. Hand reamers are made with either straight flutes (Figure 10-1 and Figure 10-2) or helical flutes (Figure 10-3 and Figure 10-4).

Figure 10-1   Hand reamer with straight flutes.
(Courtesy National Twist Drill & Tool Company.)

The hand reamer is used to give the hole a smooth finish and correct diameter. Taper reamers are available both in Morse taper

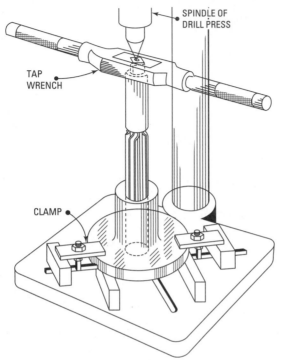

Figure 10-2 Using a drill press to steady a hand reamer. Note the lathe center in the drill press.

Figure 10-3 Hand reamer with helical flutes.
(Courtesy National Twist Drill & Tool Company.)

(Figure 10-5) and in Brown and Sharpe taper (Figure 10-6) for hand finishing.

Other hand reamers are the *adjustable hand reamer* (Figure 10-7) and the *expansion hand reamer* (Figure 10-8). The expansion reamer with the helical flute is suited for use in holes where the cut is interrupted by a longitudinal slot or keyway. *Taper-pin reamers* (Figure 10-9 and Figure 10-10) in both straight and spiral flutes are available for hand operations.

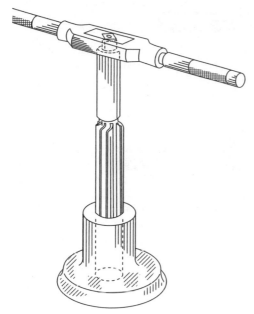

Figure 10-4 Hand reamer used to clean up the inside of this part.

Figure 10-5 Taper reamer for hand-finishing Morse-taper holes in sockets, sleeves, and spindles. *(Courtesy Morse Twist Drill & Tool Company.)*

Figure 10-6 Taper reamer for hand reaming Brown and Sharpe taper holes. *(Courtesy Morse Twist Drill & Tool Company.)*

Figure 10-7 Adjustable hand reamer. *(Courtesy Morse Twist Drill & Tool Company.)*

Figure 10-8 Expansion hand reamer with helical flutes. Especially suited for use in holes where the cut is interrupted by a longitudinal slot or keyway. *(Courtesy Morse Twist Drill & Tool Company.)*

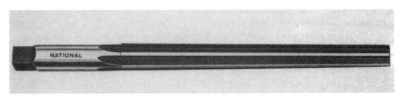

Figure 10-9 Taper-pin reamer with straight flutes. Taper is ¼ inch per foot. *(Courtesy National Twist Drill & Tool Company.)*

Figure 10-10 Taper-pin reamer with spiral flutes. Taper is ¼ inch per foot. *(Courtesy National Twist Drill & Tool Company.)*

If possible, all hand reaming should be done in a vertical position. The hand reamer should not be expected to remove a considerable amount of stock. The largest amount of stock to expect the hand reamer to remove is 0.005 inch. Holes should be drilled with this in mind when the hand reamer is to be used. These reamers are slightly tapered at the end for a distance of ⅜ inch to ½ inch, and they are 0.010 inch to 0.012 inch smaller at the end. This taper facilitates the entrance of the reamer and enables it to make a good start.

Machine Reamers

The beginning students should be able to recognize the difference between hand and machine reamers because the hand reamer will be ruined if it is used in a machine. The machine reamer has a tang and tapered shank, like a twist drill. It is inserted in a sleeve for use in the drill press spindle. Machine reamers are sometimes used before hand reamers. Several types of machine reamers are available, including the following:

- Stub screw-machine reamers
- Chucking reamers

- Expansion chucking reamers
- Rose chucking reamers
- Shell reamers
- Tapered reamers

Stub Screw-Machine Reamers

These reamers are free-cutting production tools, economical to use because their short length practically eliminates breakage (Figure 10-11). Stub reamers are particularly desirable on production jobs where close tolerances must be maintained without losing time in gaging small parts, sharpening tools, and making machine adjustments. They are manufactured in decimal sizes to provide for accurate production where close tolerances are desired. Stub reamers are designed primarily for use in automatic screw machines where short tools are required; a pinhole through the shank is provided to adapt them for use in floating holders.

Figure 10-11 Stub screw-machine reamer. Available in decimal sizes from 0.0600 to 1.010 inch inclusive. Regularly furnished with right-hand cut and 7° left-hand helix. *(Courtesy Morse Twist Drill & Tool Company.)*

Chucking Reamers

Chucking reamers may have either straight or taper shanks, and either straight or helical flutes (Figure 10-12 and Figure 10-13).

Figure 10-12 Straight-shank, chucking reamer with straight flutes. *(Courtesy Morse Twist Drill & Tool Company.)*

Figure 10-13 Taper-shank chucking reamer with helical flutes. *(Courtesy Morse Twist Drill & Tool Company.)*

Expansion Chucking Reamers

The difference between the expansion chucking reamer and the expansion hand reamer is that the expansion chucking reamer (Figure 10-14) should not be used as an adjustable reamer to cut oversize holes. This reamer is adapted to light finishing cuts in all types of materials.

Figure 10-14 Expansion chucking reamer for light finishing cuts in all types of materials. It should not be used to cut oversize holes.
(Courtesy Morse Twist Drill & Tool Company.)

Rose Chucking Reamers

Rose chucking reamers are made primarily for roughing cuts (Figure 10-15). The leading edge of the rose reamer is given a back-off chamfer of 45°. The reamer gets its name from the fact that the end view has a rose-like appearance.

Figure 10-15 Rose chucking reamer. Used primarily for roughing cuts.
(Courtesy Morse Twist Drill & Machine Company.)

Shell Reamers

The shell reamer is used for sizing and finishing operations. It is a reamer head that fits either a straight-shank or a taper-shank arbor. Several different sizes of shell reamers may be fitted to the same arbor, which saves parts.

Shell reamers are available in either straight flutes or in helical flutes (Figure 10-16 and Figure 10-17). These reamers are also available with either straight shanks or tapered shanks for machine use. The roughing reamers have lateral grooves across the lands, while the finishing reamers have straight, tapered lands. The fluted

Figure 10-16 Straight-fluted shell reamer. *(Courtesy Morse Twist Drill & Machine Company.)*

Figure 10-17 Helical-fluted shell reamer. *(Courtesy Morse Twist Drill & Machine Company.)*

shell reamer is made to remove small amounts of metal, and cuts along the edges. The *rose shell reamer* has no clearance along the lands and is made to cut on the end. It is usually a few thousandths of an inch under size so that the hole can be finish-reamed with either a hand reamer or a fluted chucking reamer. The rose shell reamer is not intended to be a finishing reamer.

A *shell-type expansion chucking reamer* with replaceable shells (Figure 10-18) is designed for high-precision reaming in the mass production of parts that have close-tolerance holes. The expansion feature is designed to compensate for loss of size caused by wear.

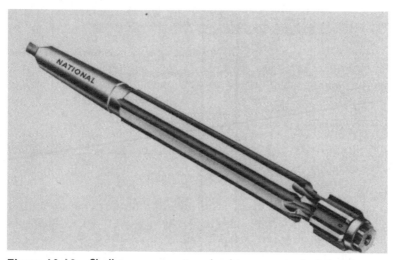

Figure 10-18 Shell-type expansion chucking reamer designed for high-precision reaming of parts that have close-tolerance holes.
(Courtesy National Twist Drill & Tool Company.)

Carbide-tipped reamers are used in many production setups and especially where abrasive material such as foundry sand and scale are encountered. The carbide-tipped, helical-fluted chucking reamer (Figure 10-19) is recommended for highly abrasive materials, heat-treated steels, and other hand materials.

Figure 10-19 Helical-fluted chucking-reamer taper shank. *(Courtesy National Twist Drill & Tool Company.)*

Tapered Reamers

Several types of tapered reamers are available for use both in hand reaming and in machine reaming operations.

Taper-pin reamers (Figure 10-20 and Figure 10-21) are all of the same taper; the point of each reamer will enter the hole reamed by the next smaller size. A *high-spiral machine, taper-pin reamer* (Figure 10-22) is designed especially for machine reaming of taper pinholes. This reamer can be run at high speeds, and chips do not clog in the flutes because of their spiral construction.

Figure 10-20 Taper-pin reamer. A hand reamer with straight flutes.
(Courtesy Morse Twist Drill & Machine Company.)

Figure 10-21 Taper-pin reamer. A hand reamer with spiral flutes.
(Courtesy Morse Twist Drill & Machine Company.)

Figure 10-22 Taper-pin reamer. A machine reamer with high spirals.
(Courtesy Morse Twist Drill & Machine Company.)

Morse-taper reamers are available for both hand reaming and for machine reaming of standard Morse-taper holes in sleeves, sockets, and spindles. A machine Morse taper is shown in Figure 10-23.

Figure 10-23 Morse-taper reamer for machine finishing of Morse-taper holes. *(Courtesy Morse Twist Drill & Machine Company.)*

Reamers of the tapered type used for special purposes are the *bridge reamer* (Figure 10-24) and the *car reamer* (Figure 10-25). The bridge reamer is used for reaming the rivet and bolt holes in structural iron and steel, boilerplates, and similar work. The deep flutes provide adequate space for chips sheared out by the reamer, and the tapered point facilitates entering holes that are out of alignment.

Figure 10-24   Bridge reamer. Used for reaming the rivet and bolt holes in structural iron and steel. *(Courtesy Morse Twist Drill & Machine Company.)*

Figure 10-25   Car reamer. *(Courtesy Morse Twist Drill & Machine Company.)*

The *repairman's taper reamer* (Figure 10-26) is used by automobile and bicycle repairmen, blacksmiths, electricians, machinists, plumbers, and carpenters for enlarging holes in thin metals, etc. The straight-shank repairman's taper reamer (Figure 10-27) may be used in an electric drill.

Figure 10-26   Repairman's taper reamer. Used for enlarging holes in thin metals. *(Courtesy Greenfield Tap & Die.)*

Figure 10-27   Repairman's taper reamer with straight shank for use in an electric drill. *(Courtesy Greenfield Tap & Die.)*

The *diemaker's reamer* (Figure 10-28) is another special-purpose reamer. In laying out a die, holes can be drilled close together,

Figure 10-28 Diemaker's reamer. Taper is 0.013 inch per inch.

(Courtesy Morse Twist Drill & Machine Company.)

outlining the shape desired, then reamed with the diemaker's reamer until the holes run together and the central piece drops out. The reamer can then be run along the outline of the edge of the die as a spiral mill; the resulting clearance is correct for the finished die. Rapid action and freedom from clogging are two valuable features of this reamer. Another valuable feature is its freedom from breakage, as the easy shearing cut imposes little strain upon the cutting edges.

Pipe reamers are tapered reamers for reaming holes to be tapped with American National Standard Taper Pipe taps (Figure 10-29). The pipe reamer is also made with inserted lands (Figure 10-30). All sizes of these reamers are tapered starting ¾ inch from the foot. The nominal size refers to the pipe size for which the reamer is intended, rather than the actual size of the reamer.

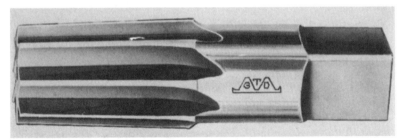

Figure 10-29 Taper pipe reamer with straight flutes for reaming holes to be tapped with American National Standard Taper Pipe taps.

(Courtesy Greenfield Tap & Die.)

Burring reamers are designed for removing the internal burrs caused by cutting pipe (Figure 10-31 and Figure 10-32). The straight-shank burring reamer may be used in an electric drill. Burring reamers are also used for countersinking and for enlarging holes in sheet metal. A *ratchet burring reamer* (Figure 10-33) may also be used for enlarging holes in sheet metal, countersinking, etc.

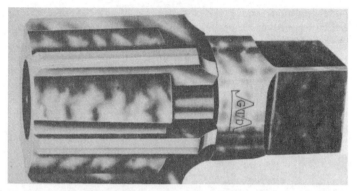

Figure 10-30   Taper pipe reamer with inserted lands. *(Courtesy Greenfield Tap & Die.)*

Figure 10-31   Burring reamer for removing interior burrs caused by cutting pipe. *(Courtesy Greenfield Tap & Die.)*

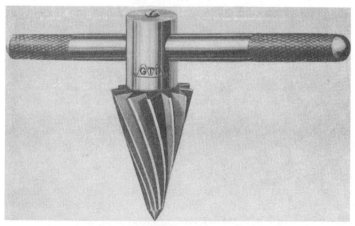

Figure 10-32   Burring reamer with T-handle shank. *(Courtesy Greenfield Tap and Die.)*

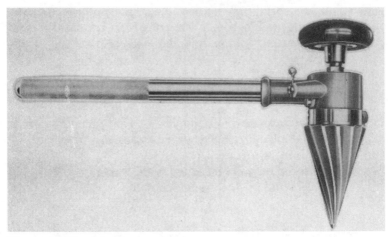

Figure 10-33 Ratchet burring reamer. *(Courtesy Greenfield Tap & Die.)*

Use and Care of Reamers

Proper use of reamers determines the accuracy of drilled holes and the quality of their finish. Proper resharpening of reamers, suitable cutting fluids, rigidity of machine tools and fixtures, and correct feeds and speeds are all factors related to the proper use of reamers.

An adequate amount of stock should be left for the reamer to cut. About 0.001 inch to 0.003 inch of stock should be allowed for hand reaming. The stock allowances should be greater for machine reaming, depending on the size: ¼-inch diameter, 0.010; ½-inch diameter, 0.015; and 1-inch diameter, 0.020.

The recommended speeds for reamers are about two-thirds the recommended speeds for twist drills. However, the feeds can be greater than for drilling. Between 0.0015 inch and 0.004 inch per flute per revolution is recommended for reamers. Chatter, oversize holes, and poor finish usually result from lack of alignment and rigidity in machine tools and fixtures.

Reamers should not be permitted to become excessively dull before resharpening. Proper equipment should be used for sharpening, and great care should be exercised.

Reamers should be transported and stored in containers that have separate compartments for each reamer. These tools are delicate and easily damaged. When not in use, they should be covered with a rust preventative.

Summary

There are many types of reamers in use. A reamer has its place in industry as a finishing and sizing tool. The use of reamers has been somewhat reduced by the introduction of the internal grinder, but work that must undergo machining in the lathe, for example, may be finished and brought to size with a reamer. Reamers may be used on work in the drill press without removing the work from the drill press, and on milling machines. Hand tools are extensively used by plumbers, electricians, and sheet metal workers.

A reamer is a precision tool designed to finish (to a given diameter) a hole that has been made by drilling. Where precision is required, the hole is first drilled to within a few thousandths of an inch of the correct size and then is reamed to the exact size.

There are various types of hand and machine reamers. Hand reamers have a square end on the shank to engage a hand wrench. The flute or cutting part on hand reamers is slightly tapered on the end to start the reamer properly. Hand reamers are made with either straight or helical flutes.

There is a difference between machine and hand reamers. If a hand reamer is used in a machine, it will be ruined. A machine reamer has a shank similar to a twist drill and is made for high-speed operation. Stub reamers are designed primarily for use in automatic screw machines where short tools are required. A pinhole through the shank is provided to adapt the reamer for use in floating holders.

Burring reamers are designed to remove the internal burrs caused by cutting pipe. Straight-shank burring reamers may be used in an electric drill. Ratchet reamers are generally used on the job by plumbers and electricians. Burring reamers are also used for countersinking and for enlarging holes in sheet metal.

Carbide-tipped reamers are used in many production setups and especially where abrasive materials such as foundry sand and scale are encountered. The carbide-tipped reamer is highly recommended for use in abrasive materials, heat-treated steels, and other hand materials.

Proper use of reamers determines the accuracy of drilled holes and the quality of their finish. Proper resharpening of reamers, suitable cutting fluids, rigidity of machine tools and fixtures, and correct feeds and speeds are all factors related to the proper use of reamers.

Review Questions

1. What is a reamer?

2. Describe a hand reamer.

3. Taper reamers are available in _____ taper and in Brown and Sharpe taper for hand finishing.

4. Where is the expansion hand reamer best suited?

5. Describe the following machine reamers:

   a. Stub screw-machine reamer

   b. Chucking reamer

   c. Expansion chucking reamer

   d. Rose chucking reamer

   e. Shell reamer

   f. Tapered reamer

   g. Carbide-tipped reamer

   h. Taper-pin reamer

   i. Morse-taper reamer

   j. Diemaker's reamer

   k. Burring reamer

6. The recommended speeds for reamers are about _____ the recommended speeds for twist drills.

7. How should reamers be stored?

8. What should be used to sharpen reamers?

9. How should reamers be transported?

10. Why shouldn't reamers be allowed to become excessively dull?

Chapter 11

Taps

A *tap* is a tool used for cutting internal threads. Taps are threaded accurately and fluted; the flutes extend the length of the threaded portion, forming a series of cutting edges. Taps are made of carbon steel and high-speed steel. Carbon steel taps are used for hand-tapping cast iron and general work that is not too severe. High-speed steel taps are used for either hand or machine tapping on tough or abrasive materials (such as aluminum, brass, Bakelite, malleable iron, die castings, fiber, hard rubber, low-carbon steels, and other materials with similar characteristics).

The threads of taps are either *cut* or *ground*. Precision work requires taps that have ground threads. Taps are available with two, three, or four flutes. The flutes may be straight, angular, or helical.

Types of Taps

The object of the tapping operation is to make helical grooves, or threads, in holes so that they may hold bolts, studs, screws, and so on. The most common taps are the hand tap and the machine-screw tap. Other types of taps are pipe, pulley, nut, taper, and special taps of various kinds.

Hand Taps

Originally, hand taps were intended only for hand operation (Figures 11-1 through 11-4), but they are now widely used on machine production work. Standard hand taps, starting with machine-screw size No. 0, are made in sets of three (Figure 11-5). These sets of three are standard in the Unified Screw Thread form as follows:

- The *taper* tap is used to start the thread (Figure 11-5A). At least six threads are tapered or chamfered before full diameter of the tap is reached.
- The *plug* tap is used to cut the threads as fast as possible after the taper tap has been used (Figure 11-5B). Three to five threads of the plug tap are chamfered or tapered.
- The *bottoming* tap is used last to drive the thread to the bottom of a blind hole (Figure 11-5C). This tap has only 1½ threads chamfered.

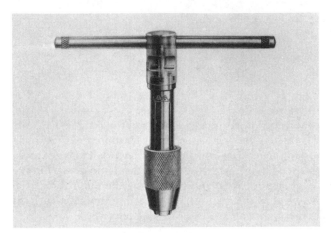

Figure 11-1 T-handle tap wrench with slip handle designed for use in hand tapping. This wrench can also be used on twist drills, reamers, screw extractors, etc. *(Courtesy Greenfield Tap & Die.)*

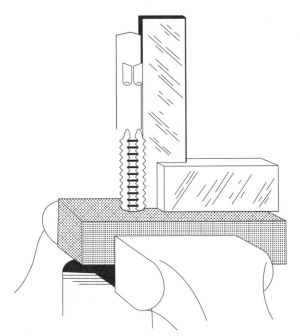

Figure 11-2 Using a try-square to align the tap in the hole for better cutting action and better internal threads.

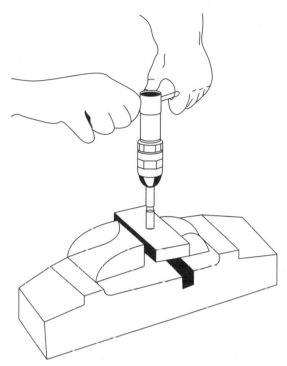

Figure 11-3   Using the T-Handle tap wrench requires both hands.

Figure 11-4   Tap wrench with solid handles. One handle is movable and can be turned to adjust the center. *(Courtesy Greenfield Tap & Die.)*

These taps are used when tapping a blind hole (a hole that does not extend entirely through the work) and it is desired to cut threads to the bottom of the hole. For a hole extending entirely through the work, of course, only the taper tap and plug tap are necessary.

Taps are available in forms other than the Unified Thread form. Other thread forms are metric, Whitworth, Acme, and square threads.

(A) Taper tap (10 threads chamfered).

(B) Plug tap (3 to 5 threads chamfered).

(C) Bottoming (1½ threads chamfered).

Figure 11-5 Set of three hand taps used in succession for tapping a blind hole. These taps are identical in general dimensions, their only difference being in the chamfered threaded portion at the point. *(Courtesy Greenfield Tap & Die.)*

Size of Taps

The size of hand taps ¼ inch in diameter and larger is indicated in fractional dimensions. Machine-screw taps are really small standard hand taps. Their size is indicated by the machine-screw system of sizes, ranging from No. 0 (0.060 inch) up to No. 12 (0.216 inch).

Both the fractional screw sizes and the machine-screw sizes are available in taper, plug, and bottoming taps in both Unified National Coarse and Unified National Fine Threads. The sizes of Unified National Coarse Standard Screw Threads are shown in Table 11-1, and the sizes of Unified National Fine Standard Screw Threads are shown in Table 11-2.

Table 11-1 Unified National Coarse Standard (UNC) Screw Thread Pitches and Recommended Tap Drill Sizes (Formerly American National Form Thread)

Sizes	Threads per Inch	Outside Diameter of Screw	Tap Drill Sizes	Decimal Equivalent of Drill
1	64	0.073	53	0.0595
2	56	0.086	50	0.0700
3	48	0.099	46	0.0810

Sizes	Threads per Inch	Outside Diameter of Screw	Tap Drill Sizes	Decimal Equivalent of Drill
4	40	0.112	43	0.0890
5	40	0.125	38	0.1015
6	32	0.138	33	0.1130
8	32	0.164	29	0.1360
10	24	0.190	25	0.1495
12	24	0.216	16	0.1770
1/4	20	0.250	7	0.2010
5/16	18	0.3125	F	0.2570
3/8	16	0.375	5/16	0.3125
7/16	14	0.4375	U	0.3680
1/2	13	0.500	27/64	0.4219
9/16	12	0.5625	31/64	0.4843
5/8	11	0.625	17/32	0.5312
3/4	10	0.750	21/32	0.6562
7/8	9	0.875	49/64	0.7656
1	8	1.000	7/8	0.875
1 1/8	7	1.125	63/64	0.9843
1 1/4	7	1.250	1 7/64	1.1093

Courtesy South Bend Lathe, Inc.

Table 11-2 · Unified National Fine Standard (UNF) Screw Thread Pitches and Recommended Tap Drill Sizes

Sizes	Threads per Inch	Outside Diameter of Screw	Tap Drill Sizes	Decimal Equivalent of Drill
0	80	0.060	3/64	0.0469
1	72	0.073	53	0.0595
2	64	0.086	49	0.0730
3	56	0.099	44	0.0860
4	48	0.112	42	0.0935
5	44	0.125	37	0.1040
6	40	0.138	32	0.1160
8	36	0.164	29	0.1360
10	32	0.190	21	0.1590
12	28	0.216	14	0.1820
1/4	28	0.250	7/32	0.2187

(continued)

Table 11-2 *(continued)*

Sizes	Threads per Inch	Outside Diameter of Screw	Tap Drill Sizes	Decimal Equivalent of Drill
⁵⁄₁₆	24	0.3125	I	0.2720
³⁄₈	24	0.375	R	0.3390
⁷⁄₁₆	20	0.4375	²⁵⁄₆₄	0.3906
¹⁄₂	20	0.500	²⁹⁄₆₄	0.4531
⁹⁄₁₆	18	0.5625	³³⁄₆₄	0.5156
⁵⁄₈	18	0.625	³⁷⁄₆₄	0.5781
³⁄₄	16	0.750	¹¹⁄₁₆	0.6875
⁷⁄₈	14	0.875	¹³⁄₁₆	0.8125
1	14	1.000	¹⁵⁄₁₆	0.9375
1¹⁄₈	12	1.125	1³⁄₆₄	1.0468
1¹⁄₄	12	1.250	1¹¹⁄₆₄	1.1718

Courtesy South Bend Lathe, Inc.

Size of Tap Drill to Use

As tap size is determined by the major diameter of its threads, it is evident that it is necessary to drill the hole for tapping smaller than the tap size by nearly twice the depth of thread. The drilled hole must be small enough to leave sufficient stock in which to cut the screw threads. Generally, a tap drill that will give approximately a 75-percent thread is used. Tap drill size can be determined either from Table 11-1 or Table 11-2, or by use of the following formula:

$$\text{Tap drill size} = \text{Major diameter (of thread)} - \frac{(0.75 \times 1.299)}{\text{No. threads per inch}}$$

For example, for a hole ³⁄₄ inch in diameter, 10 threads per inch, substitute the following in the previous formula.

$$\text{Tap drill size} = 0.750 - \frac{(0.75 \times 1.299)}{10}$$

$$= 0.750 - 0.097$$

$$= 0.653 \text{ (decimal equivalent of size of tap drill)}$$

$$= {}^{21}\!/_{32} \text{ (correct tap drill size nearest 0.653)}$$

For all practical purposes a simpler formula can be used as follows:

$$\text{Tap drill size} = \text{Major diameter} - \frac{1}{N(\text{No. threads per inch})}$$

$$= \frac{3}{4} - \frac{1}{10}$$

$$= 0.750 - 0.1$$

$$= 0.650 \text{ or nearest decimal to } \frac{21}{32}$$

Other Types of Hand Taps

In addition to the standard taps, consisting of the taper, plug, and bottoming taps already mentioned, other styles of hand taps are available in which the points and flutes are designed differently.

Spiral-pointed or *"gun" hand taps* with plug chamfer are designed primarily for use in through holes (Figure 11-6A). The spiral point forces the chips ahead of the tap. This prevents the chips from clogging in the flutes and causing tap breakage. It also eliminates possible damaging of the threads by the chips when the tap is reversed. The bottoming chamfered spiral-pointed tap is used in blind holes where a space is provided at the bottom of the hole for chip collection (Figure 11-6B).

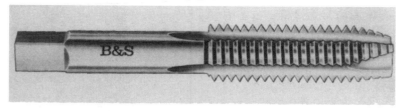

(A) Plug chamfer.

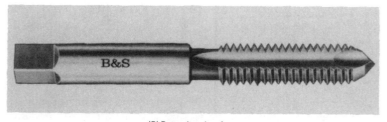

(B) Bottoming chamfer.

Figure 11-6 Spiral-pointed hand taps. *(Courtesy American Twist Drill Company.)*

Spiral-fluted hand taps are recommended for tapping blind holes in such ductile materials as aluminum and magnesium, where chip removal is a problem (Figure 11-7). These taps are most effective when the material being tapped produces long, stringy, curling chips (Figure 11-8). The taps cut freely while ejecting chips from the tapped hole, which prevents clogging and damage to both the threaded parts and the taps. *Fast spiral-fluted hand taps* are recommended for deep blind holes where chip removal is a problem (Figure 11-9).

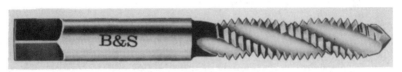

(A) Plug chamfer.

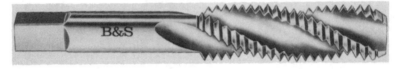

(B) Bottoming chamfer.

Figure 11-7 Spiral-fluted hand taps. *(Courtesy American Twist Drill Company.)*

Spiral-pointed hand taps with a short flute are used for through hole tapping of holes in sheet metal and other thin sections (Figure 11-10). This design is suitable where the tapped hole is not deeper than the diameter of the tap.

Machine-Screw Taps

There is really no difference in standard hand taps and machine-screw taps, except that the latter are made in the machine-screw system of sizes. In this system, No. 0 is equivalent to 0.060 inch with a regular increment of 0.013 inch between each size. Thus, No. 1 equals 0.073 inch, No. 2 equals 0.086 inch, No. 3 equals 0.099 inch, and so on. Machine-screw taps are used for tapped holes below ¼ inch in diameter. They are available in the Unified Screw form of thread in taper, plug, and bottoming styles.

Machine-screw taps are also available in several styles in which the points and flutes are designed for improved performance in certain applications and materials. The *fast spiral-fluted machine-screw tap* is shown in Figure 11-11. It is recommended for tapping deep blind holes in aluminum and magnesium.

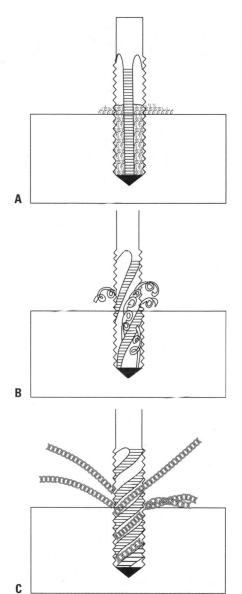

Figure 11-8 Note the cutting actions of these taps: (A) Hand tap, (B) Low-angle spiral-fluted tap, (C) High-angle spiral-fluted tap.

(Courtesy of Greenfield Tap and Die.)

A

B

C

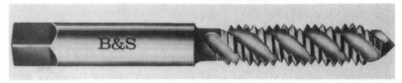

(A) Plug chamfer.

(B) Bottoming chamfer.

Figure 11-9 Fast spiral-fluted hand taps. *(Courtesy American Twist Drill Company.)*

Figure 11-10 Spiral-pointed hand tap with short flute (plug chamfer).
(Courtesy American Twist Drill Company.)

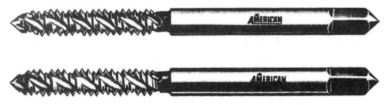

(A) Plug chamfer.   (B) Bottoming chamfer.

Figure 11-11 Fast spiral-fluted machine-screw tap.
(Courtesy American Twist Drill Company.)

Table 11-3 shows metric tap drill sizes, or tapping drill diameters for metric screw threads.

Thread forming (roll) taps (Figure 11-12) are fluteless taps that do not cut threads in the same manner as conventional taps. These taps are forming tools, and the threading action is similar to the rolling process used to produce external threads. Thread-rolling taps produce a strong thread, and, because of the forming action, the thread surface is somewhat work-hardened. Tap drills for roll taps must be larger than those used for the same diameter thread with conventional taps.

Table 11-3   Metric Tap Drill Sizes—Tapping Drill Diameters for Metric Screw Threads

Nominal Diameter of Thread (mm)	Depth of External Thread (mm)	Tapping Drill Diameter (mm)	Radial Engagement with External Thread (%)
1.6	0.2147	1.30	70
1.8	0.2147	1.50	70
2	0.2454	1.65	71.3
2.2	0.2760	1.80	72.5
2.5	0.2760	2.10	72.5
3	0.3067	2.60	65.2
3.5	0.3681	3.00	68
4	0.4294	3.40	70
4.5	0.4601	3.90	65.3
5	0.4908	4.30	71.3
6	0.6134	5.20	65.4
7	0.6134	6.20	65.4
8	0.7668	7.00	65.2
9	0.7668	8.00	65.2
10	0.9202	8.80	65.1
11	0.9202	9.80	65.1
12	1.0735	10.60	65.2
14	1.2269	12.40	65.2
16	1.2269	14.25	71
18	1.5336	16.00	65.2
20	1.5336	18.00	65.2
22	1.5336	20.00	65.2
24	1.8403	21.50	68
27	1.8403	24.50	68
30	2.1470	27.00	69.8

Pipe Taps

Pipe taps are made in two thread forms: straight and taper. The nominal size of a pipe tap is the same as that of the fitting to be tapped—not the actual size of the tap. *Straight pipe taps* (Figure 11-13) with American Standard Pipe Thread (NPS) form are intended for tapping holes or couplings for low-pressure work to assemble with taper-threaded pipe or fittings to secure a tight joint with the use of a lubricant or sealer.

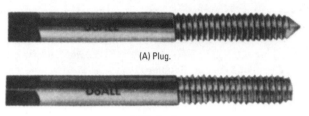

(A) Plug.

(B) Bottoming.

Figure 11-12 Thread-forming (roll) taps. *(Courtesy DoAll Company.)*

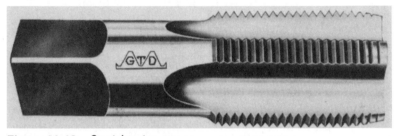

Figure 11-13 Straight pipe tap. *(Courtesy Greenfield Tap & Die.)*

Taper-pipe taps are available with both regular and interrupted threads (Figure 11-14 and Figure 11-15). These taps are furnished with American Standard Pipe (NPT) form of thread having a taper of ³⁄₄ inch per foot. The interrupted-thread taps have every other thread removed, except for a few threads at the point. These taps are used for tapping tough metals that have a tendency to "load" the teeth of the tap and should be used only when the regular thread taps fail.

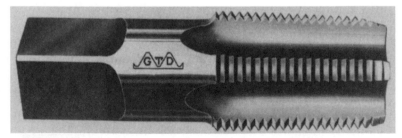

Figure 11-14 Taper-pipe tap with regular thread. *(Courtesy Greenfield Tap & Die.)*

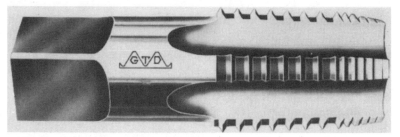

Figure 11-15 Taper-pipe tap with interrupted thread. Every other thread is removed except for the first few threads at the point.
(Courtesy Greenfield Tap & Die.)

Nut Taps

These taps are used where the work requires a tap with a long shank, or where the hole is of a greater depth than can be handled by a hand tap (Figure 11-16). Nut taps are used for tapping small quantities made from tough material, such as stainless steel and similar alloys.

Figure 11-16 Nut taps. *(Courtesy Greenfield Tap & Die.)*

Pulley Taps

The setscrew and oilcup holes in the hubs of pulleys are tapped with pulley taps (Figure 11-17). These taps are available in several different overall lengths because of the variation in diameter of pulleys. The long shank also permits tapping in places that might be inaccessible to hand taps.

Figure 11-17 Pulley taps used for tapping setscrew and oilcup holes in the hubs of pulleys. *(Courtesy Greenfield Tap & Die.)*

Taper Taps

Taper taps are used by nut manufacturers. They are regularly furnished with long shanks and short threads, and as succeeding nuts are tapped, they run up on the shank until it is full; then the tap is removed from the holder, and the nuts slide off (Figure 11-18). The

Figure 11-18 Taper tap with straight shank. *(Courtesy Greenfield Tap & Die.)*

threaded section is made as short as practical, and a smaller number of teeth are chamfered on the nut taps.

Bent-shank taper taps (Figure 11-19) are designed for use in automatic tapping machines manufactured by the National Machinery Company. Tapping in these machines is continuous—the nuts are fed to the tap automatically. There are different sizes of the machine; each machine requires a given shank length and radius in the bend.

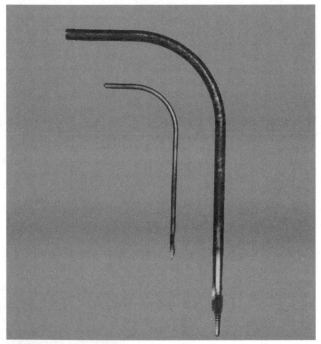

Figure 11-19 Bent-shank taper tap. *(Courtesy Greenfield Tap & Die.)*

Special-Purpose Taps
Spark-plug taps are available with the International form of thread in plug style only.

Acme threads are used extensively for transmitting and controlling power in lead screws on machine tools, and on valves, jacks, and other mechanisms. Acme tap sets usually consist of one or more roughing taps and a finishing tap. The ground-thread taps are usually used on Acme screw and nut assemblies, because there is less wear when the lead of the screw and nut are matched accurately.

Tap Selection

A review of the principles involved may be helpful in tap selection. The following tap features must be considered in selecting the proper tap for a given job:

- The type of tap
- Carbon steel or high-speed steel tap
- Cut thread or ground thread
- Thread limits—If a precision or ground-thread tap is to be used

Generally, the type of tap is determined by the nature of the job and by the tapping facilities available. A plant equipped with standard general-purpose tapping machines may find the proper tap to be one of the so-called hand taps, depending on the nature of the job (i.e., the depth of the hole, cutting characteristics of the material being tapped, and so on). A nut manufacturer equipped with machines that use bent-shank taper taps has no choice in the type of tap, but is concerned only with its size. In some instances, particular conditions or requirements may require that a special tap be made to order.

Special taps include any size, style, or finish of taps not included in standard listings in the manufacturer's catalogue. Special recommendations can be made only when the exact conditions are known because the type of tap is so dependent on factors beyond the control of the tap manufacturers.

Speed, properties of the material to be tapped, and desired accuracy are all important factors to be considered in deciding whether to use carbon steel or high-speed steel taps. Carbon steel taps may be run about one-half as fast as high-speed steel taps.

Carbon steel taps may be used efficiently in brass and ferrous metals. Most nonferrous metals and abrasive materials (such as Bakelite, fiber, or hard rubber) turn cutting edges quickly and require the use of high-speed steel taps. If accuracy is required, the ground-thread taps necessary for this type of service are regularly available only in high-speed steel.

Accuracy is the only important factor to be considered when determining whether to use cut-thread or ground-thread taps. Cut-thread taps are formed by milling or "cutting." The manufacturing tolerances or variations in size and form of cut threads are not held to such uniformity or to such close "limits" as are the ground-thread taps. Ground-thread taps are finish ground with abrasive wheels to extremely close tolerances. Accuracy of the required thread is determined by the class of thread desired in the threaded parts. Thus, if a tapped hole and a threaded stud are required to "fit" to a very close tolerance, *threads* of "closer limits" will be required.

To select correctly the size of tap for a particular job, an understanding of the relationship between *tap pitch diameter tolerance* and *gage limits* is necessary. Pitch diameter tolerances are indicated as follows:

- For taps through 1 inch diameter

  L1 = Basic to Basic minus 0.0005 inch

  H1 = Basic to Basic plus 0.0005 inch

  H2 = Basic plus 0.0005 inch to Basic plus 0.001 inch

  H3 = Basic plus 0.001 inch to Basic plus 0.0015 inch

  H4 = Basic plus 0.0015 inch to Basic plus 0.002 inch

  H5 = Basic plus 0.002 inch to Basic plus 0.0025 inch

  H6 = Basic plus 0.0025 inch to Basic plus 0.003 inch

- For taps over 1 inch diameter through 1½ inch diameter

  H4 = Basic plus 0.001 inch to Basic plus 0.002 inch

Classes of Thread

The various degrees of snugness of fit are expressed in terms of class of thread. The classes of thread are provided in Table 11-4.

Table 11-4   Classes of Thread—Unified Thread

Class	Description
Class 1A and 1B	The combination of Class 1A for external threads and Class 1B for internal threads is intended to cover the manufacture of threaded parts *where quick and easy assembly is necessary or desired* and an allowance is required to permit ready assembly.

Table 11-4 *(continued)*

Class	Description
Class 2A and 2B	The combination of Class 2A for external threads and Class 2B for internal threads is designed *for screws, bolts, and nuts.* It is also suitable for a wide variety of other applications. A similar allowance is provided that minimizes galling and seizure encountered in assembly and use. It also accommodates, to a limited extent, platings, finishes, or coatings.
Class 3A and 3B	The combination of Class 3A for external threads and Class 3B for internal threads is provided for those applications *where closeness of fit and accuracy of lead and angle of thread are important.* These threads are obtained consistently only by use of high-quality production equipment supported by a very efficient system of gaging and inspection. No allowance is provided.

Courtesy American Twist Drill Company.

The six classes include three for screws and three for nuts. The external classes of threads are identified as 1A, 2A, and 3A; 1B, 2B, and 3B refer to the internal classes of threads. Class 2A and 2B threads are the free-fitting type found on a majority of the commercial threaded fasteners. Any unified class of external thread may be mated with any internal class of thread as long as the product meets the specified tolerance and allowance.

After the class of threads is determined, the standard gage limits of threads that correspond to that fit must be considered, and a tap selected to produce threads conforming to these predetermined limits of size. If the "limits" of cut-thread taps are not close enough to ensure the class of fit or quality of thread desired, the choice, of course, is a ground-thread tap.

Determination of either too fast or too slow tapping speed is essential to efficient tapping. There are certain speeds at which taps operate efficiently in specific materials, and these are shown in Table 11-5. The composition of the material to be threaded, the kind of steel from which the tap is made, and the design of both the tap and tapping machine are all important factors in determining the proper tapping speed for a given job. High or maximum speeds may be determined by gradual stages of experimentation.

**Table 11-5   Recommended Cutting Speeds and Lubricants
for Machine Tapping**

Material	Speeds in Feet per Minute (ft/min)	Lubricant
Aluminum	90–100	Kerosene and light-base oil
Brass	90–100	Soluble oil or light-base oil
Cast iron	70–80	Dry or soluble oil
Magnesium	20–50	Light-base oil diluted with kerosene
Phosphor bronze	30–60	Mineral oil or light-base oil
Plastics	50–70	Dry or air jet
Steels		
Low carbon	40–60	Sulfur-base oil
High carbon	25–35	Sulfur-base oil
Free machining	60–80	Sulfur-base oil
Molybdenum	10–35	Soluble oil
Stainless	10–35	Soluble oil

Summary

A tap is a precision tool used for cutting internal threads. The threads of taps are either cut or ground. Precision work requires taps that have ground threads. Taps are available with two, three, or four flutes, which may be straight, angular, or helical. Hand taps were intended only for hand operation, but are widely used on machine production work. Standard hand taps are generally made in sets of three: taper tap, plug tap, and bottoming tap. The taper tap is used to start the thread, the plug tap is used to cut the threads as far as possible, and the bottoming tap is used to drive the thread to the bottom of the hole. These taps are used when tapping a blind hole (a hole that does not extend entirely through the work).

Taps are available in forms other than Unified Thread form. Other thread forms are metric, Whitworth, Acme, and square threads.

Other types of hand taps are spiral-point, which are used generally in through holes, and spiral-fluted, which are recommended for tapping blind holes in ductile materials such as aluminum and magnesium.

Determination of either too fast or too slow tapping speed is essential to efficient tapping. There are certain speeds at which taps operate efficiently in specific materials, which are shown in a table of values. The composition of the material to be threaded, the kind of steel from which the tap is made, and the design of both the tap and tapping machine are all important factors in determining the proper tapping speed for a given job. High or maximum speeds may be determined by gradual stages of experimentation.

Review Questions

1. How many flutes are generally on a tap?
2. Can hand taps be used in machine work?
3. Standard taps are generally made in sets of three. Name them.
4. What must be considered when selecting a tap for a given job?
5. How much stock should be left for threads in a drilled hole?
6. What is the object of the tapping operation?
7. For what is the taper tap used?
8. For what is the plug tap used?
9. How is the bottoming tap used?
10. How is the size of hand taps ¼ inch in diameter and larger indicated?
11. List types of hand taps *other than* the taper, plug, and bottoming taps.
12. How do fluteless taps cut threads?
13. Where do you use a nut tap?
14. Where do you use bent-shank taper taps?
15. What four tap features do you need to know to select a tap properly?

Chapter 12

Threading Dies

A threading die is a tool with an internal thread (similar to the thread in a nut) that is used to cut *external threads* on bolts and round stock. The thread is cut by the teeth of the die as the die advances, while being turned or screwed onto round stock.

Types of Dies

Several types of thread-cutting dies are available to meet varied conditions. The various threads were discussed in Chapter 11.

Solid Dies

A special holder is not needed for solid dies. A large wrench may be used to turn them. Solid dies are used principally for rethreading bruised or rusty threads, but they may also be used for cutting new threads (Figure 12-1 and Figure 12-2).

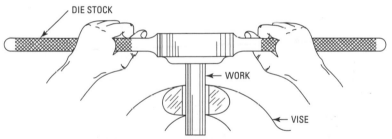

Figure 12-1   Starting to cut threads.

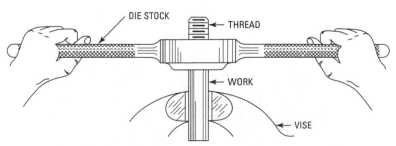

Figure 12-2   Cutting threads.

Figure 12-3 Solid square bolt dies. *(Courtesy Greenfield Tap & Die.)*

Solid square bolt dies are one-piece nonadjustable dies (Figure 12-3). They are used for cutting threads on bolts and pipes, as well as for repair work on threads.

Solid hexagon rethreading dies, as the name implies, are used principally for repairing bruised or rusty threads (Figure 12-4). They can be turned with any size wrench that will fit.

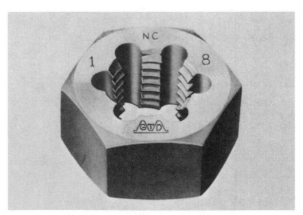

Figure 12-4 Hexagon rethreading dies. *(Courtesy Greenfield Tap & Die.)*

Round-Split Dies

Adjustable dies of the split type are made in all standardized thread sizes, have limited adjustment for size, and cut threads in easy stages. The round-split die is split on one side and can be adjusted to the desired class of thread. Most split dies are set to cut the thread either slightly over or slightly under the designated size.

Round-split dies are of two types: *screw adjusting* and *open adjusting*. Adjustment of the screw-adjusting type is by means of a fine-pitch screw, which forces the sides of the split die apart, or allows them to spring together (Figure 12-5). This adjustment remains positive when the die is removed from the machine holder or from the hand stock, so that each time the die is used, a new adjustment is not necessary. The slot in these dies is beveled so that the adjusting screw can be removed

Figure 12-5 Screw-adjusting type of round adjustable die. A fine-pitch screw forces the sides of the die apart or allows them to spring back together.

(Courtesy Morse Twist Drill & Machine Company.)

(if necessary) when the dies are used in a machine holder and adjustment can be made by the adjusting screw in the holder. The open type of die (Figure 12-6), designed for use in both screw machines and diestocks, is adjusted by means of three screws in the holder, one screw for expanding and two screws for compressing the dies (Figure 12-7).

Figure 12-6 The open type of round die. *(Courtesy Greenfield Tap & Die.)*

Figure 12-7 The three-screw type of round diestock for holding round adjustable dies. *(Courtesy Greenfield Tap & Die.)*

Since the range of adjustment of round dies is limited, only slight adjustments are possible. Adjustments several thousandths of an inch larger than the nominal size of the die result in poor performance because of drag on the heel of the thread sections. Excessive expansion may cause the die to break in two. If it is necessary to cut a thread that requires more than a slight adjustment of the die, a two-piece die should be used.

Two-Piece Adjustable Dies

This type of adjustable die is made in two separate halves (Figure 12-8). The halves are usually held in a collet, which consists of a cap and a guide (Figure 12-9). The die is adjusted by setscrews at either end of the slot. Pipe threading dies may also be of the two-piece type. These are regularly available with the standard pipe taper of $\frac{3}{4}$ inch per foot.

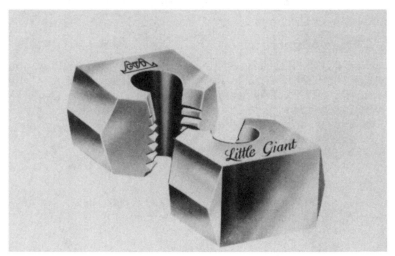

Figure 12-8 Two-piece adjustable die. *(Courtesy Greenfield Tap & Die.)*

Acorn Dies

Acorn dies are used in specially designed holders. The Acorn die has a tapered nose at the end of the lands, accurately ground concentric with the thread (Figure 12-10). A solid adjusting cap (Figure 12-11) effects a uniform closing in of the lands when it is tightened. The die is driven by lugs on the holder that fit two slots in its base. The adjustment is governed by screwing the cap more or less tightly onto the holder. Proper adjustment is maintained by a lock nut.

(A) Cap.   (B) Guide.

(C) Collet.

Figure 12-9   Collet for use with two-piece dies, consisting of a cap and a guide. *(Courtesy Greenfield Tap & Die.)*

Figure 12-10   Acorn dies for use in specially designed holder.
(Courtesy Greenfield Tap & Die.)

Figure 12-11   Die-holder reducing cap enables the Acorn die holder to hold dies of a small diameter. *(Courtesy Greenfield Tap & Die.)*

Acorn die holders (Figure 12-12) are designed for use on automatic screws and other machines that provide for automatically reversing the die, or the rod, at the instant the desired length of thread has been cut. The body has a longitudinal "float" that allows the die to follow its own lead independently of any lag in the machine.

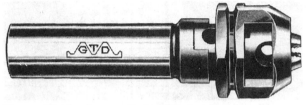

Figure 12-12   Acorn die holder designed for use on automatic machines. *(Courtesy Greenfield Tap & Die.)*

Use of Dies to Cut Threads

A diestock with two-piece adjustable dies is used to cut pipe threads. The die and ways should be absolutely clean of chips, dust, and so on before placing the die halves in the ways of the diestock. Adjust the two-piece die by loosening the holding bolts, and turn each adjusting bolt until the reference mark on each half die registers with the mark "S" on the diestock (Figure 12-13). When these are in position, tighten the holding bolts firmly. Use only the wrench provided with the stock for this purpose.

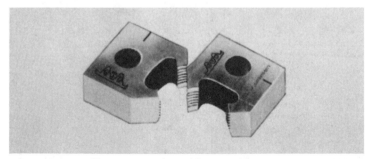

Figure 12-13   Two-piece adjustable pipe dies. *(Courtesy Greenfield Tap & Die.)*

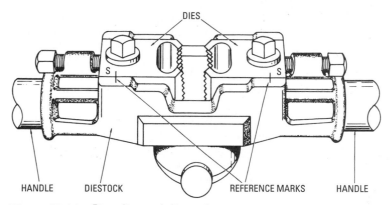

Figure 12-14   Pipe dies and diestock.

The die will cut a standard thread when the reference marks register properly (Figure 12-14). Adjusting is done only when irregularity or variations in fittings (elbows, Ts, etc.) make it necessary. After the die halves are properly placed and all adjustments are made, select the proper guide collar, place it in the sleeve, and secure it by tightening the thumb bolt. Screw the two arms into place, and the tool is ready to cut threads. The pipe to be threaded should be clamped in a pipe vise (Figure 12-15).

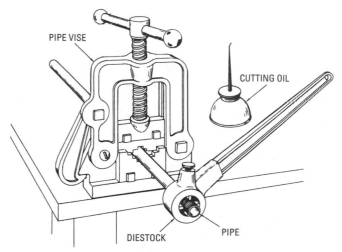

Figure 12-15   Pipe threading.

Use plenty of cutting oil when starting and cutting the thread. In starting, press the dies firmly against the pipe end until they "take hold." After a few turns, blow out the chips and apply more oil. This should be repeated two or three times before completing the cut. After the thread cutting is completed, blow out the chips and back off the die. Avoid frequent reversals, which are made by most pipefitters.

Soluble oil has been found to be preferable to lubricating oil when used as a cutting oil. The heat generated when cutting threads is dissipated by the water in the emulsion that flows to the cutting edge of the die, giving continuous lubrication, rather than spasmodic flooding, as with lubricating oil.

Summary

Threading dies are tools with internal threads similar to the threads on a nut. These tools are used to cut threads on a bolt or almost any round material (such as steel). The threads are cut by the teeth of the die as the die advances while being turned or screwed onto round stock.

There are various types of dies, such as solid, round-split, two-piece adjustable, and Acorn. The solid die is one piece and nonadjustable and is used for cutting threads on standard bolts and pipe. This type of die is also used to repair bruised or rusty threads. The round-split die is split on one side and can be closed or opened gradually to the desired size of the thread. Most split dies are set to cut the thread slightly under the designated size.

Solid dies are used primarily for rethreading bruised or rusted threads, but may also be used for cutting new threads. A special holder is not needed. Use a large wrench.

Two-piece adjustable dies are made in two separate halves and are usually held together in a collet, which consists of a cap and a guide. The die is adjusted by setscrews at either end of the slot. Acorn dies are used in specially designed holders and have a tapered nose at the end of the lands accurately ground concentric with the thread. Acorn die holders are designed for use on automatic screws and other machines that provide for automatically reversing the die.

Soluble oil has been found to be preferable to lubricating oil as a cutting oil. The heat generated when cutting threads is dissipated by water in the emulsion that flows to the cutting edge of the die, giving continuous lubrication, rather than spasmodic flooding (as with lubricating oils).

Review Questions

1. Name the four types or designs of threading dies.
2. How are the two-piece adjustable dies adjusted?
3. Why are cutting fluids used?
4. What tool is used to hold a die?
5. How much can a split die be adjusted?
6. What is a threading die?
7. Identify the following types of dies:
   a. Solid dies
   b. Round-split dies
   c. Two-piece adjustable dies
   d. Acorn dies
8. A diestock with two-piece adjustable dies is used to cut _____ threads.
9. When using a die to cut threads it is a good idea to use plenty of _____ oil.
10. Name the four types or designs of threading dies.
11. How are the two-piece adjustable dies adjusted?
12. Why are cutting fluids used in cutting threads?
13. What tool is used to hold a die?
14. How much can a split die be adjusted?
15. Soluble oil has been found to be preferable to _____ oil as a cutting oil.

Chapter 13

Milling-Machine Cutters

Milling cutters are usually referred to as multitooth, cylindrical, rotary cutting tools designed for mounting on milling-machine arbors. The principal dimensions of the most commonly used types of milling cutters have been adopted as standard by the cutting manufacturers, and these standards have been approved by the American Standards Association. The most commonly used cutters in milling operations are made of high-speed steel, cemented carbide, or cast alloys.

Milling Operation

For many years, *up milling*—rotating the cutter opposite the direction of feed (Figure 13-1)—was considered the only practical way to use milling cutters. In recent years, however, *down milling*—rotating the cutter in the direction of feed (Figure 13-2)—has been recognized.

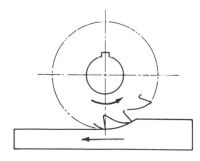

Figure 13-1 Up-milling action.
(Courtesy National Twist Drill and Tool Company.)

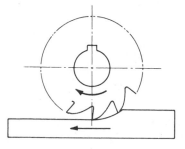

Figure 13-2 Down-milling action.
(Courtesy National Twist Drill and Tool Company.)

In the up-milling operation, the cutter tooth has a tendency to slide along the surface for a short distance. This sliding action under pressure tends to dull the cutter tooth. The cutter revolution marks so familiar on milled surfaces are caused by the alternate sliding action and breaking through of the cutter teeth. Down milling is not practical on all milling machines. This method should not be used unless the nature of the job permits that both the work and the cutter be held rigidly, and the milling machine is equipped with an anti-backlash device. If down milling can be used, however, a better surface finish, larger feeds per tooth, and longer cutter life without regrinding can be expected.

In down milling, full engagement of the tooth with the work occurs practically instantaneously. Thus, the gradual building up of peripheral pressures and the resulting sliding action and dulling of the cutter are prevented. Also, gradual disengagement of the teeth with the work largely eliminates feed marks.

Classification of Milling Cutters

Milling cutters may be classified by clearance (or relief) of teeth. *Profile cutters* are sharpened by grinding on the periphery of the teeth. The clearance is obtained by grinding a narrow land back of the cutting edge. The cutters are called *shaped-profile cutters* if the cutting edges are curved or irregular in shape. *Formed cutters* are sharpened by grinding the face of the teeth. The *eccentric relief* (or clearance) back of the cutting edge has the same contour as the cutting edge itself.

Another classification of cutters is based on the method of mounting. Cutters with a hole for mounting on an arbor are designated *arbor cutters*. Cutters having either a straight shank or a tapered shank integral with the cutter are called *shank cutters*. Those cutters that can be attached directly to a spindle end or a stub arbor are called *facing cutters*.

Milling cutters are either *right-hand* or *left-hand* cutters. A right-hand cutter rotates counterclockwise, and a left-hand cutter rotates clockwise, as viewed from the front when mounted on the spindle.

General Types of Milling Cutters

The milling-machine operator should be familiar with each cutter by name and with the operation that it can perform. The different milling cutters are designed for a specific purpose.

Plain Milling Cutters

The *plain milling cutter* (commonly called a *mill*) is used to mill flat surfaces parallel to the axis of rotation. The cutter teeth have a $12\frac{1}{2}°$ rake. Cutters with less than a $\frac{3}{4}$-inch face have straight (axial) teeth, and the larger sizes have left-hand helical or "spiral" teeth. The left-hand helical tooth causes the cutting thrust, which tends to keep the spindle tight in its bearings.

Plain mills are adapted for work of a slabbing nature, when the work is narrower than the cutter face. When used on flat work where there is a shoulder, the lead end of helical teeth should work in the corner, and the shallow-end teeth (or side chip grooves) should be ground into that end for best results.

Light-duty plain cutters are best suited for moderate cuts in malleable iron, steel, and cast iron (Figure 13-3). Heavy-duty plain cutters have a large heavy rake, coarse teeth, deep flutes, and a steep helix (Figure 13-4).

Figure 13-3 Plain milling cutter used for light-duty milling.

(Courtesy National Twist Drill and Tool Company.)

Figure 13-4 Plain milling cutter used for heavy-duty milling.

(Courtesy National Twist Drill and Tool Company.)

A *helical cutter* is a plain milling cutter with an extra-steep helical angle, usually 52° (Figure 13-5). The helix is generally opposite the direction of rotation, thereby utilizing the end thrust to keep the spindle tight in its bearings. This is not a general-purpose cutter. It can be run with a light cut at high speeds and fast feeds on either brass or soft steels. It cuts with a shearing action, forces the chips off sidewise, does not show revolution marks, and does not spring away from the work. This feature makes it especially adaptable for use on thin work or for intermittent cuts where the amount of stock to be removed varies. For extremely wide surfaces, cutters can be made in sections with the spiral angle reversed in each succeeding section.

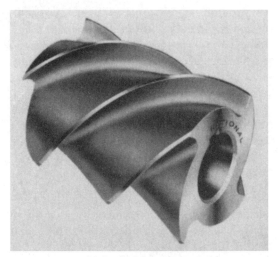

Figure 13-5 Helical plain milling gear. *(Courtesy National Twist Drill and Tool Company.)*

The term *spiral* should not be used interchangeably for helix. A *helix* is a curve generated by a point that both rotates and advances axially on a cylindrical surface. A *spiral* is a curve generated by a point having three motions: (1) rotation about an axis, (2) increase in distance from the axis, and (3) advancement parallel to the axis. Plain cutters with a helix angle of 25° to 45° are commonly, but incorrectly, called *spiral mills*.

Coarse-tooth cutters are capable of removing a considerable quantity of metal in a given time without overloading the cutter or machine. The wide spaces between the teeth permit the cutting edges to be well backed up, which is not always possible with closely spaced teeth. Therefore, the cutters are well adapted to handle deep and rapid cuts without danger of failing.

Nicked milling cutters are made with nicked or grooved teeth to enable the cutter to take deeper cuts (such as for roughing work). The effect of the nicks is to reduce the power required to drive the cutter. The nicks are arranged so that a cutting edge (of the next tooth) will be behind a nick. Thus, instead of a continuous chip, a number of chips are made by each cutting edge. Nicked cutters that have a very wide (long) face are known as slabbing cutters.

Side Milling Cutters

These cutters are plain milling cutters of cylindrical form that have teeth around the periphery and on one or both sides (Figure 13-6). Side mills are recommended for side milling, for slotting, and for straddle milling work. If the cutter has teeth on only one side, it is a *half-side milling cutter* (Figure 13-7). Half-side mills are used for heavy-duty straddle mill work.

The *staggered-tooth side milling cutter* (Figure 13-8) is designed for deep slotting and for heavy-duty side milling. The shear cutting action (alternately right and left) eliminates side thrust. The alternate right- and left-hand spirals of the teeth, with considerable angle of undercut, enable this cutter to remove a large amount of metal without destructive vibration and chatter, permit taking deep cuts, and leave a good finish. Free cutting action makes increased speed and feed possible. Cuts that would stall an ordinary

Figure 13-6 Side milling cutter used for slotting and light-duty milling. Cutters can be ganged and used as straddle mills. *(Courtesy National Twist Drill and Tool Company.)*

cutter can be taken easily. Although they were intended primarily for deep cuts in steel, staggered-tooth side mills can be used for shallow cuts, which is an advantage if the work requires cuts of varying depths. Ganging two or more of these cutters, instead of using a wide cutter with a wide tooth space, is recommended when wide slots are to be milled.

Interlocking side cutters are useful where a slot width must be held to extremely accurate limits. These cutters can be separated by

Figure 13-7 Half-side milling cutters used for heavy-duty straddle milling. *(Courtesy National Twist Drill and Tool Company.)*

Figure 13-8 Staggered-tooth side milling cutter used for deep slotting and for heavy-duty side milling. Cutters can be ganged and used as straddle mills. *(Courtesy National Twist Drill and Tool Company.)*

spacing collars of the required thickness to obtain the correct width of face.

The shearing action (alternately right and left) eliminates side thrust; the cutting action is very smooth and rapid. The cutters can also be made with inserted teeth, in the larger sizes. The teeth have a positive rake on all cutting faces. These cutters have the added advantage of being suitable for finishing the bottom of a slot, while having an adjustment for holding the slot to a required width.

Inserted-tooth cutters (Figure 13-9) are commonly used for larger cutters. Inserted teeth (sometimes called blades) are generally used because this construction is cheaper, and all teeth can be replaced easily if necessary. Inserted-tooth construction avoids the danger of cracking while being hardened. The teeth may be made of either high-speed steel, cemented carbide, or cast alloy.

Figure 13-9 Milling cutter with solid carbide indexible throwaway inserts. *(Courtesy National Twist Drill and Tool Company.)*

Various methods of holding the inserted teeth are used. They are generally made long enough to permit sharpening a great number of times. These cutters are used for heavy-duty side milling and for face-milling jobs where long life under severe working conditions is desirable.

End Mills

An *end mill*, by strict definition, is a milling cutter that has cutting teeth only on its end. However, in addition to the end teeth, end mills may have teeth along the periphery or cylindrical surface.

Because they have cutting teeth on the end of the mill, end mills are usually held by shanks. They may have either a straight shank (Figure 13-10) or a taper shank (Figure 13-11) to fit various collets and adapters.

Figure 13-10   Straight-shank end mill used for general-purpose end-milling operations. *(Courtesy Morse Twist Drill & Machine Company.)*

Figure 13-11   Taper-shank end mill. *(Courtesy Morse Twist Drill & Tool Company.)*

End mills may be made for either right-hand or left-hand rotation. The helix (right-hand or left-hand) may be in either the same or the opposite direction as the cutter rotation. When the helix and cutter rotation are the same (either right-hand or left-hand), the teeth have a positive rake angle. For some specific purposes, end mills are available with the cutter rotation and the helix running in opposite directions (for example, left-hand cutter rotation with right-hand helix). Generally, end mills that have both a right-hand helix and a right-hand cutter rotation are preferred.

Several types of cutting ends for end mills are produced to further adapt them to a wide variety of uses (such as profiling, end milling, slotting, surface milling, and many other milling-machine operations). The different types of cutting ends available on end mills are:

- Two-flute
- Multiple-flute (three-flute, four-flute, six-flute)
- Single-end
- Double-end
- Hollow (solid and adjustable)
- Ball-end
- Carbide-tipped
- Shell-end

Two-flute, single-end end mills are adapted to slotting operations in all kinds of materials. This mill can cut to center, which permits plunge cutting (Figure 13-12).

Figure 13-12 Two-flute, single-end end mill.
(Courtesy Morse Twist Drill & Machine Company.)

Figure 13-13 Two-flute, double-end end mill.
(Courtesy Morse Twist Drill & Machine Company.)

A *two-flute, double-end end mill* is shown in Figure 13-13. It is also adapted for slotting operations in all types of materials.

The *adjustable-type* and the *solid-type hollow mills* are shown in Figure 13-14. Hollow mills are used for sizing bar stock of all types in screw machines or in turret lathes. They are available with undercut teeth for use in steel or with straight teeth for use in brass. Compensation for internal wear can be made on the adjustable type of hollow mill.

(A) Adjustable type.

(B) Solid type.

Figure 13-14 Hollow mills used in screw machines or turret lathes for sizing bar stock of all types. *(Courtesy Morse Twist Drill & Machine Company.)*

Ball-end end mills are used for machining fillets and slots with corner radii. They are also used extensively for die sinking and machining dies (Figure 13-15).

Figure 13-15 Four-flute, ball-end end mill. *(Courtesy National Twist Drill and Tool Company.)*

Carbide-tipped end mills are available in most types and shapes (Figure 13-16). The advantage of carbide-tipped cutters is increased cutting speeds. The surface feet per minute for carbide-tipped cutters is double that of high-speed steel.

Figure 13-16   Carbide-tipped, straight-flute end mill. *(Courtesy National Twist Drill and Tool Company.)*

Shell-end end mills are larger than solid end mills and range from 1¼ inches to 6 inches in diameter. These mills have a hole in the center to mount the cutter on an arbor (Figure 13-17). Cutters of this type are used for slabbing or surface cuts.

Figure 13-17   High-speed, steel-shell mill. *(Courtesy National Twist Drill and Tool Company.)*

Angle Milling Cutters

Angle milling cutters are designed to mill at an angle to the axis of rotation. They are used for milling surfaces at various angles to the axis of rotation and are often used in making other milling cutters.

Angle cutters are made for right-hand rotation and for left-hand rotation. *Single-angle milling cutters* (Figure 13-18) are used to mill ratchet teeth or to mill dovetails. The common single-angle cutters vary from 40° to 80°. *Double-angle milling cutters* (Figure 13-19) are available with included angles of 45°, 60°, or 90°.

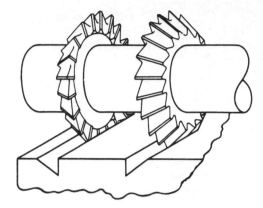

Figure 13-18 Single-angle milling cutter used for milling ratchet teeth or for milling dovetails. *(Courtesy Morse Twist Drill & Machine Company.)*

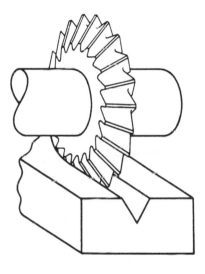

Figure 13-19 Double-angle milling cutter used for milling grooves, notches, serrations, or threads. *(Courtesy Morse Twist Drill & Machine Company.)*

Single-angle cutters have one side at an angle of 90° to the axis of rotation, and the other side at, usually, either 45° or 60°—only one side cuts at an angle other than 90° to the rotation axis. A double-angle cutter is constructed in such a manner that two angles cut at an angle other than 90° to the rotation axis.

Slitting Saws, Slotting Saws, and Miscellaneous Cutters

Thin, straight-toothed, plain milling cutters are generally called *slitting saws*. A *plain metal-slitting saw* is shown in Figure 13-20. This saw is used for general-purpose slotting, parting, or cutting-off

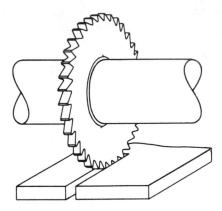

Figure 13-20 Plain metal-slitting saw. *(Courtesy National Twist Drill & Machine Company.)*

operations of moderate depth in both ferrous and nonferrous materials.

Side-tooth metal-slitting saws have side chip-clearance gashes (Figure 13-21). They are used for cutting off operations in all types of materials.

The *staggered-tooth metal-slitting saw* (Figure 13-22) is manufactured with alternate helical teeth for shear cutting action and chip clearance between the side teeth. These saws are used for heavy-duty slotting in all types of materials and can make deeper cuts under coarser feeds than other types of saws.

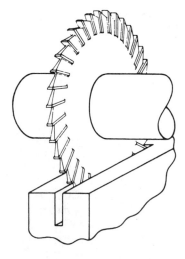

Figure 13-21 Side-tooth metal-slitting saw. Note the chip-clearance gashes on the side. *(Courtesy Morse Twist Drill & Machine Company.)*

A *screw-slotting cutter* is shown in Figure 13-23. This cutter is ground with side clearance and is used for slotting both ferrous and nonferrous screwheads, sheet, or tubing. Screw-slotting cutters are used only for making shallow, short slots similar to those in screwheads.

A *T-slot milling cutter* is a special form of end mill for making T-slots (Figure 13-24). This cutter is designed to mill the wide bottom of a T-slot after the narrow portion has been milled with a side mill or an end mill.

Woodruff key-seat cutters are made in both shank and arbor types. The shank type (Figure 13-25) is used for cutting the smaller Woodruff key seats. The larger Woodruff key seats are cut with the arbor type (Figure 13-26) of key-seat cutter. These cutters have profile teeth and are used for cutting semicircular keyways in shafts.

The *threaded-hole, single-angle milling cutter* (Figure 13-27) may be used for dovetail milling in all types of materials. The cutter with a 60° included angle is generally used for milling dovetails.

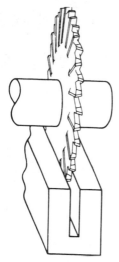

Figure 13-22
Staggered-tooth metal-slitting saw used for heavy-duty slotting in all types of materials. *(Courtesy Morse Twist Drill & Machine Company.)*

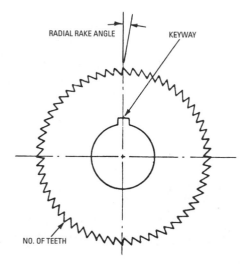

RADIAL RAKE ANGLE KEYWAY

NO. OF TEETH

Figure 13-23 Screw-slotting cutter. Used for slotting ferrous and nonferrous screwheads, sheet, or tubing. *(Courtesy National Twist Drill and Tool Company.)*

Figure 13-24 T-slot milling cutter. *(Courtesy Morse Twist Drill & Machine Company.)*

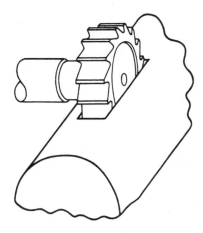

Figure 13-25 Woodruff key-seat cutter (shank type). Used to cut the smaller Woodruff key seats.

(Courtesy Morse Twist Drill & Machine Company.)

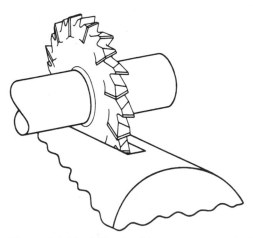

Figure 13-26 Woodruff key-seat cutter (arbor type). Used to cut the larger Woodruff key seats. *(Courtesy Morse Twist Drill & Machine Company.)*

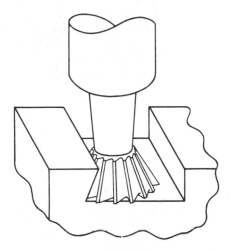

Figure 13-27 Threaded-hole, single-angle cutter used for dovetail milling. *(Courtesy Morse Twist Drill & Machine Company.)*

Form-Relieved Cutters

Form-relieved cutters are special milling cutters shaped to meet various milling specifications. This cutter is sharpened by grinding only the faces of the teeth. The outline of the teeth corresponds to the required form, and should not be ground when resharpening.

Convex cutters (Figure 13-28) are used to mill half-circles in all types of materials. If convex half-circles are desired, a *concave cutter* (Figure 13-29) must be used. Convex quarter-circles are milled with *corner-rounding cutters* (Figure 13-30).

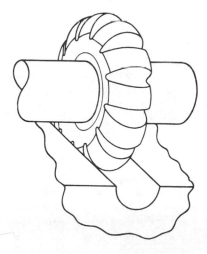

Figure 13-28 Convex cutter. A form-relieved cutter used to mill concave half-circles in all types of materials. *(Courtesy Morse Twist Drill & Machine Company.)*

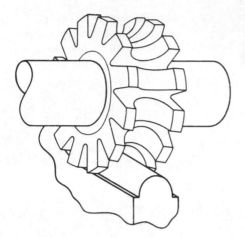

Figure 13-29 Concave cutter. A form-relieved cutter used to mill convex half-circles in all types of materials. *(Courtesy Morse Twist Drill & Machine Company.)*

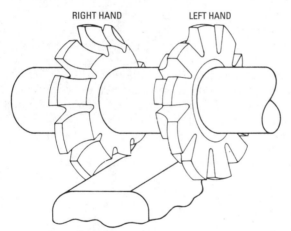

RIGHT HAND   LEFT HAND

Figure 13-30 Corner-rounding cutter. A form-relieved cutter used for milling convex quarter-circles on the edges of all types of materials. *(Courtesy Morse Twist Drill & Machine Company.)*

Gear cutters are available in many different diameters and hole sizes, so that they can be used on almost any regular milling machine or arbor. *Finishing-gear milling cutters* (Figure 13-31) with a 14½°-pressure-angle, regular-involute form are most commonly used. These cutters can be sharpened on the face of the teeth without changing their form. Finishing-gear milling cutters are made with eight cutters for each pitch—eight cutters being required to produce a full set of gears, ranging from 12 teeth to a rack.

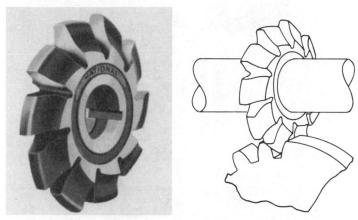

Figure 13-31   Finishing gear milling cutter. A form-relieved cutter with 14½° pressure angle. *(Courtesy National Twist Drill and Tool Company.)*

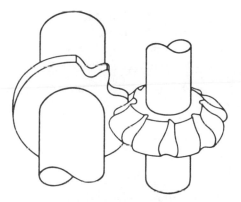

Figure 13-32   Sprocket cutter. Used for cutting roller chain sprockets.

(Courtesy Morse Twist Drill & Machine Company.)

A *sprocket cutter* is shown in Figure 13-32. This cutter is used for cutting roller chain sprockets that have the American National Standard Roller Chain Tooth Form.

Hobs

A *hob* is a hardened, threaded cutter, formed like a worm (Figure 13-33). It is commonly used for cutting teeth in spur and helical gears, herringbone gears, worm gears, worms, splined shafts, ratchets, square shafts, and sprockets for silent chain, roller chain, and block chain.

Figure 13-33 Hob. Used to hob spur gears and helical gears.
(Courtesy National Twist Drill and Tool Company.)

Hobs are made in either the ground or unground form. For average hobbing operations, unground forms are more economical and more commonly used. Ground-form hobs are better for extreme accuracy.

Spur-gear hobs can be used to hob either spur gears or helical gears. A right-hand hob can be used to cut either a right-hand helical gear or a left-hand helical gear. However, it is recommended that a right-hand hob be used to cut a right-hand helical gear, and a left-hand hob be used to cut a left-hand helical gear.

In general, the hobbing process can be employed to produce any form that is repeated regularly on the periphery of a circular part.

Care of Milling Cutters

Every precaution should be taken to prevent milling cutters from becoming nicked and dulled. In making milling-machine setups, the cutters should not be bumped against the workpiece, the tools, or the machine. It is common practice to coat the cutting edges of cutters with plastic to prevent nicking and dulling. The cutters should be stored carefully when they are not in use.

Speeds and Feeds

Individual experience and judgment are extremely valuable in selecting the correct milling speeds and feeds. Even though suggested rate tables are given, it should be remembered that these are only suggestions. The lower figure in the table for a particular material should always be used until sufficient practical experience has been gained. Then, the speed can be increased until either excessive cutter wear or chatter indicates that the practical limit has been exceeded.

Speed and feed rates are governed by many variable factors. Some of these are material, cutter, width and depth of cut, required surface finish, machine rigidity and setup, power and speed available, and cutting fluid.

Speeds

The speed at which the circumference of the cutter passes over the work is always given in surface feet per minute (sfpm). However, the spindle speed of a milling machine is always given in revolutions per minute (rpm). Table 13-1 can be used to convert surface feet per minute (sfpm) to revolutions per minute (rpm), for making speed adjustments on the milling machine. The table gives the rpm for cutters of different diameters at various surface speeds. Speeds not listed in the table can be found by simple calculation—for example, for 200 sfpm, 1¾-inch-diameter cutting must run twice as fast as for 100 sfpm ($2 \times 509 = 1018$). If a table of speeds is unavailable, a formula can be used for spindle speed as follows:

$$r/min = \frac{sf/min \times 12}{\pi \times diameter}$$

The cutting speeds for the different materials to be milled are given in Table 13-2. These speeds are based on average conditions for high-speed steel tools. The cutting speeds may be doubled for carbide-tipped cutters. Note that the cutting speeds in Table 13-2 are given in surface feet per minute (sfpm). If a table of cutting speeds is unavailable, sfpm may be calculated by the following formula:

$$sf/min = \frac{\pi \times diameter \times r/min}{12}$$

Milling cutters with helical teeth can be run faster than those with straight teeth because the shearing action enables them to cut more freely. Coarse-tooth cutters can be run faster than fine-tooth cutters because fewer teeth are in contact with the work at any time. Roughing cuts are usually made at slow speeds and heavy feeds; finishing cuts are made at high speeds and fine feeds. However, there are exceptions to these practices.

Table 13-1 Revolutions per Minute

Diameter, in.	Surface Feet per Minute						
	40	**50**	**60**	**70**	**80**	**90**	**100**
¼	611	764	917	1070	1222	1375	1528
⁵⁄₁₆	489	611	733	856	978	1100	1222
³⁄₈	407	509	611	713	815	917	1019
⁷⁄₁₆	349	437	524	611	698	786	873
½	306	382	458	535	611	688	764
⁵⁄₈	244	306	367	428	489	550	611
¾	204	255	306	357	407	458	509
⁷⁄₈	175	218	262	306	349	393	437
1	153	191	229	267	306	344	382
1⅛	136	170	204	238	272	306	340
1¼	122	153	183	214	244	275	306
1⅜	111	139	167	194	222	250	278
1½	102	127	153	178	204	229	255
1⅝	94	117	141	165	188	212	235
1¾	87	109	131	153	175	196	218
1⅞	81	102	122	143	163	182	204
2	76	95	115	134	153	172	191
2¼	68	85	102	119	136	153	170
2½	61	76	92	107	122	137	152
2¾	52	69	83	97	111	125	139
3	51	64	76	89	102	115	127
3½	44	55	65	76	87	98	108
4	38	48	57	67	76	86	95
4½	34	42	51	59	68	77	85
5	31	38	46	54	61	69	76
5½	28	35	42	49	56	63	70
6	25	32	38	45	51	57	64
7	22	27	33	38	44	49	55
8	19	24	29	33	38	43	48
9	17	21	25	30	34	38	42
10	15	19	23	27	31	34	38
11	14	17	21	24	28	31	35
12	13	16	19	22	25	29	32
13	12	15	18	21	24	27	29
16	10	12	14	17	19	22	24
18	8	11	13	15	17	19	21

Courtesy Cincinnati Milcron Inc.

Table 13-2   Cutting Speeds (Surface Feet per Minute)

Material	High-Speed Steel		Carbide-Tipped		Coolant
	Rough	Finish	Rough	Finish	
Cast iron	50–60	80–110	180–200	350–400	Dry
Semi-steel	40–50	65–90	140–160	250–300	Dry
Malleable iron	80–100	110–130	250–300	400–500	Soluble, sulfurized, or mineral oil
Cast steel	45–60	70–90	150–180	200–250	Soluble, sulfurized, mineral, or mineral lard oil
Copper	100–150	150–200	600	1000	Soluble, sulfurized or mineral lard oil
Brass	200–300	200–300	600–1000	600–1000	Dry
Bronze	100–150	150–180	600	1000	Soluble, sulfurized, or mineral lard oil
Aluminum	400	700	800	1000	Soluble or sulfurized oil, mineral oil and kerosene
Magnesium	600–800	1000–1500	1000–1500	1000–1500	Dry, kerosene, mineral lard oil
SAE Steels 1020 (coarse feed)	60–80	60–80	300	300	Soluble, sulfurized, mineral, or mineral lard oil
1020 (fine feed)	100–120	100–120	450	450	Soluble, sulfurized, mineral, or mineral lard oil
1035	75–90	90–120	250	250	Soluble, sulfurized, mineral, or mineral lard oil

(continued)

Table 13-2 *(continued)*

Material	High-Speed Steel		Carbide-Tipped		Coolant
	Rough	Finish	Rough	Finish	
x-1315	175–200	175–200	400–500	400–500	Soluble, sulfurized, mineral, or mineral lard oil
1050	60–80	100	200	200	Soluble, sulfurized, mineral, or mineral lard oil
2315	90–110	90–110	300	300	Soluble, sulfurized, mineral, or mineral lard oil
3150	50–60	70–90	200	200	Soluble, sulfurized, mineral, or mineral lard oil
4340	40–50	60–70	200	200	Sulfurized or mineral oil
Stainless steel	100–120	100–120	240–300	240–300	Sulfurized or mineral oil

Note: Feeds should be as great as the work and equipment will stand, provided a satisfactory surface finish is obtained.

Feeds

The *feed rate* is the rate at which the work advances past the cutter. Feed rate is commonly given in inches per minute (in/min). Generally, the rule in production work is to use all the feed that the machine and the work can stand. However, it is a problem to know where start with the feed. Tables 13-3 and 13-4 give the suggested speed per tooth for high-speed steel and carbide-tipped milling cutters, respectively.

Note that the feeds are given in thousandths of an inch per tooth for the various cutters. Multiply the feed per tooth by the number of teeth, and multiply that product by the rpm to determine the feed rate in in/min as follows:

$$\text{in/min} = \text{feed per tooth} \times \text{number of teeth} \times \text{r/min}$$

Table 13-3 Suggested Feed per Tooth for High-Speed Steel Milling Cutters

Material	Face Mills	Helical Mills	Slotting and Side Mills	End Mills	Form-Relieved Cutters	Circular Saws
Plastics	0.013	0.010	0.008	0.007	0.004	0.003
Magnesium and alloys	0.022	0.018	0.013	0.011	0.007	0.005
Aluminum and alloys	0.022	0.018	0.013	0.011	0.007	0.005
Free-cutting brasses and bronzes	0.022	0.018	0.013	0.011	0.007	0.005
Medium brasses and bronzes	0.014	0.011	0.008	0.007	0.004	0.003
Hard brasses and bronzes	0.009	0.007	0.006	0.005	0.003	0.002
Copper	0.012	0.010	0.007	0.006	0.004	0.003
Cast iron, soft (150–180 B.H.)	0.016	0.013	0.009	0.008	0.005	0.004
Cast iron, medium (180–220 B.H.)	0.013	0.010	0.007	0.007	0.004	0.003
Cast iron, hard (220–300 B.H.)	0.011	0.008	0.006	0.006	0.003	0.003
Malleable iron	0.012	0.010	0.007	0.006	0.004	0.003
Cast steel	0.012	0.010	0.007	0.006	0.004	0.003
Low-carbon steel, free machining	0.012	0.010	0.007	0.006	0.004	0.003
Low-carbon steel	0.010	0.008	0.006	0.005	0.003	0.003
Medium-carbon steel	0.010	0.008	0.006	0.005	0.003	0.003
Alloy steel, annealed (180–220 B.H.)	0.008	0.007	0.005	0.004	0.003	0.002
Alloy steel, tough (220–300 B.H.)	0.006	0.005	0.004	0.003	0.002	0.002
Alloy steel, hard (300–400 B.H.)	0.004	0.003	0.003	0.002	0.002	0.001

(continued)

Table 13-3 *(continued)*

Material	Face Mills	Helical Mills	Slotting and Side Mills	End Mills	Form-Relieved Cutters	Circular Saws
Stainless steels, free machining	0.010	0.008	0.006	0.005	0.003	0.002
Stainless steels	0.006	0.005	0.004	0.003	0.002	0.002
Monel metals	0.008	0.007	0.005	0.004	0.003	0.002

Courtesy Cincinnati Milcron Inc.

Table 13-4 Suggested Feed per Tooth for Carbide-Tipped Cutters

Material	Face Mills	Helical Mills	Slotting and Side Mills	End Mills	Form-Relieved Cutters	Circular Saws
Plastics	0.015	0.012	0.009	0.007	0.005	0.004
Magnesium and alloys	0.020	0.016	0.012	0.010	0.006	0.005
Aluminum and alloys	0.020	0.016	0.012	0.010	0.006	0.005
Free-cutting brasses and bronzes	0.020	0.016	0.012	0.010	0.006	0.005
Medium brasses and bronzes	0.012	0.010	0.007	0.006	0.004	0.003
Hard brasses and bronzes	0.010	0.008	0.006	0.005	0.003	0.003
Copper	0.012	0.009	0.007	0.006	0.004	0.003
Cast iron, soft (150–180 B.H.)	0.020	0.016	0.012	0.010	0.006	0.005
Cast iron, medium (180–220 B.H.)	0.016	0.013	0.010	0.008	0.005	0.004
Cast iron, hard (220–300 B.H.)	0.012	0.010	0.007	0.006	0.004	0.003
Malleable iron	0.014	0.011	0.008	0.007	0.004	0.004
Cast steel	0.014	0.011	0.008	0.007	0.005	0.004
Low-carbon steel, free machining	0.016	0.013	0.009	0.008	0.005	0.004

Table 13-4 (continued)

Material	Face Mills	Helical Mills	Slotting and Side Mills	End Mills	Form-Relieved Cutters	Circular Saws
Low-carbon steel	0.014	0.011	0.008	0.007	0.004	0.004
Medium-carbon steel	0.014	0.011	0.008	0.007	0.004	0.004
Alloy steel, annealed (180–220 B.H.)	0.014	0.011	0.008	0.007	0.004	0.004
Alloy steel, tough (220–300 B.H.)	0.012	0.010	0.007	0.006	0.004	0.003
Alloy steel, hard (300–400 B.H.)	0.010	0.008	0.006	0.005	0.003	0.003
Stainless steels, free machining	0.014	0.011	0.008	0.007	0.004	0.004
Stainless steels	0.010	0.008	0.006	0.005	0.003	0.003
Monel metals	0.010	0.008	0.006	0.005	0.003	0.003

Courtesy Cincinnati Milcron Inc.

In actual practice, it is better to start the feed rate at a somewhat lower figure than that indicated in the table and work up gradually until the most efficient removal rates are reached. Too high a feed is indicated by excessive cutter wear. A cutter may be spoiled by too fine a feed or too heavy a feed. Rubbing, rather than a cutting action, may dull the cutting edge, and excessive heat may be generated. This fact is also true in relationship to depth of cut. The first cut on castings and rough forgings should always be made well below the surface skin. Cuts less than 0.015 inches in depth should be avoided. To obtain a good finish, take a roughing cut followed by a finishing cut, with a higher speed and lighter feed for the finishing cut.

In general, feeds are increased as speeds are reduced. Therefore, the feed is increased on abrasive, sandy, or scaly material, and for heavy cuts in heavy work. Feed should be increased if cutter wear is excessive or if there is chatter.

Feed should be decreased for better finish, when taking deep slotting cuts, or if work cannot be held rigidly. If the cutter begins to chip or to produce long, continuous chips, the feed should be decreased.

Summary

For many years, up milling-machine cutters rotated opposite the feed. In recent years, milling-machine cutters were designed to rotate in the same direction as the feed. This type of design has eliminated the need of frequent cutter sharpening and overall maintenance. This type of cutting action is called down milling.

An end mill is a milling cutter with cutting teeth on only one end. End mills may have either right-hand or left-hand rotation. The helix may be in either the same or opposite direction as the cutter rotation. Generally, end mills that have both a right-hand helix and a right-hand cutter rotation are preferred.

Several types of cutting ends for end mills are produced. The different types of cutting ends available are two-flute, multiple-flute, single- and double-end, hollow-end, and both solid and shell.

Various shapes and sizes of cutters are produced, such as angle cutters that are generally made for right-hand rotation. The common single-angle cutters include angles of 45°, 60°, or 90°. Other shapes include double-angle, plain, side-tooth, convex, concave, and corner-rounding.

Individual experience and judgment are extremely valuable in selecting the correct milling speeds and feeds. Even though suggested rate tables are given, it should be remembered that these are only suggestions. Speed and feed rates are governed by many variable factors, including material, cutter, width and depth of cut required, surface finish, machine rigidity and setup, power and speed available, and cutting fluid.

The speed at which the circumference of the cutter passes over the work is always given in surface feet per minute (sfpm). However, the spindle speed of a milling machine is always given in revolutions per minute (rpms). Tables are available to provide conversion, and simple calculations can be made to ensure the right speed and feed.

Review Questions

1. What is the down-milling action?
2. Define the term *end milling*.
3. What is side milling?
4. Name a few cutting ends available for end milling.
5. What is the advantage of using carbide-tipped cutters?
6. How do you sharpen profile cutters?
7. How do you sharpen formed cutters?

8. What type of cutter do you use for moderate cuts in malleable iron, steel, or cast iron?

9. How are nicked milling cutters made?

10. What sizes are standard for shell-end mills?

11. What is a Woodruff key-seat cutter?

12. Where would you use a convex cutter?

13. What is a hob?

14. Explain how a sprocket cutter actually does its job.

15. How do you designate the cutter speed in relation to its circumference?

Chapter 14

Milling-Machine Arbors, Collets, and Adapters

Milling-machine operators should be familiar with the holding devices used on the machines that they operate. An inexperienced operator may mistake an *arbor* (which is used to hold milling cutters) for a *mandrel* (which is used to hold bored parts while turning the outside surface on a lathe).

Arbors

An arbor is a tapered cylindrical shaft designed to hold milling cutters. The holding part of an arbor is cylindrical. The cutter is clamped with a nut for light work and secured further by a key for heavy cutting (Figure 14-1). The precision and trueness of arbors influence the accuracy of milling operations. Arbors must be carefully handled both in use and in storage.

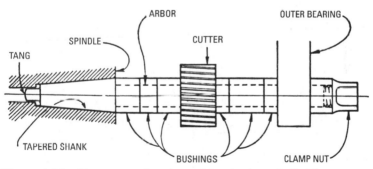

Figure 14-1 Tang-type arbor in position in the spindle. The cutter and bushings are held in a firm frictional grip by tightening the clamp nut.

Manufacturers have attempted to standardize milling machine spindles and arbors. A uniform numbering system for standard tapered arbors has been approved, and the No. 50 taper is used on most milling machines. Specifications are listed in sequence as follows: taper size, diameter, style, length from shoulder to nut, and size of bearing. For example, Arbor No. 50-1A18-4 indicates that the arbor has No. 50 standard taper, a 1-inch diameter, style

A, a length of 18 inches from the shoulder to the nut, and a No. 4 bearing.

Arbors are held firmly in the hollow spindle of the milling machine by means of an *arbor draw-in bar.* Two drive keys on the spindle nose (which fits into corresponding slots on the arbor flanges) give driving contact and positive drive to the arbor.

Styles of Arbors

Steel taper-shank arbors to fit the national milling machine standard spindle are made in three styles as follows:

- *Style A* has a pilot on the outer end (Figure 14-2). The arbor is firmly supported as it turns in the arbor support bearing, suspended from the overarm.

- *Style B* does not have a pilot (Figure 14-3). A bearing sleeve fits over the arbor and is keyed to it. The sleeve revolves in the bearing in the arbor support. Support can be placed close to the cutters on the arbor for rigidity. Style B arbors are used wherever heavy cuts are made.

- *Style C* is a short arbor requiring no arbor support (Figure 14-4). It is used to hold cutters that are too small to be bolted directly to the spindle nose, such as the smaller sizes of shell-end–mill, and face-milling cutters. This style of arbor is sometimes called a *shell-end–mill arbor.* The cutter is driven by solid lugs on the outer end of the arbor.

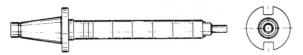

Figure 14-2   Typical milling-machine arbor, *Style A.* Note the pilot, which turns in the bearing in the arbor support. *(Courtesy Cincinnati Milacron Company.)*

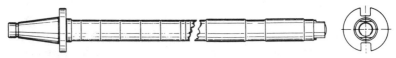

Figure 14-3   Typical milling-machine arbor, *Style B.* These arbors are longer and require support near the cutters for rigidity.
(Courtesy Cincinnati Milacron Company.)

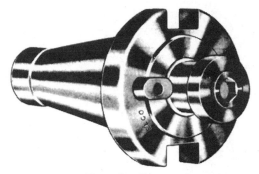

Figure 14-4 Typical milling-machine arbor, *Style C*. Used for shell-type milling cutters. *(Courtesy Brown & Sharpe Manufacturing Company.)*

Methods of Driving the Cutters

Several methods are used to prevent the cutter from slipping or turning on the arbor. The type of work to be performed and the type of cutter influence the design of the arbor.

Friction drive can be used on light work. The cutter is kept from slipping by tightening the end nut with a wrench. This forces the bushings against the cutter, and its endwise movement is prevented by the shank collar. Thus, a firm frictional grip on the cutter is produced. This is by no means a positive drive (Figure 14-5).

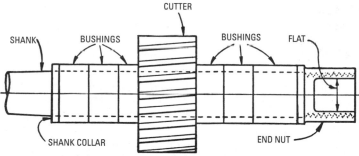

Figure 14-5 Arbor with assembled cutter and bushings, illustrating frictional cutter drive.

A *key drive* is a positive drive and prevents slipping on any work within the capacity of the machine (Figure 14-6). Cutters used on milling machines have machined keyways; a square key fits into the keyways of both the cutter and the arbor. Cutters can be mounted in any position along the arbor.

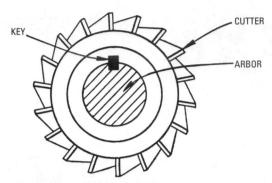

Figure 14-6 A key-cutter drive. The cutter will not slip on the arbor for any work within the capacity of the machine.

Figure 14-7 Arbor sleeve for a milling machine. *(Courtesy Brown & Sharpe Manufacturing Company.)*

Figure 14-8 Spacing collar used on a milling-machine arbor. *(Courtesy Brown & Sharpe Manufacturing Company.)*

The *arbor sleeve* has a keyway, and it may be keyed to the arbor (Figure 14-7). The sleeve fits the bushing of the arbor support. It is used as near as possible to the cutter, or cutters, on long arbors. The arbor sleeve looks like an arbor spacing collar (Figure 14-8), but it is different both in purpose and in diameter.

The *clutch drive* is used to drive gutters of the shell-end–mill type. The shell-end mill is designed with a slot on each side that fits over the arbor jaws when the mill is placed on the arbor, thus forming a positive drive. Two lugs at 180° fit in the slots of the cutter to form the driving members. A tanged-end type of arbor is shown in Figure 14-9. An arbor with a threaded end for right-hand shell-end mills only is shown in Figure 14-10.

Figure 14-9 A shell-end–mill arbor with tanged end. *(Courtesy Morse Twist Drill & Machine Company.)*

Figure 14-10 A shell-end–mill arbor with threaded end. *(Courtesy Morse Twist Drill & Machine Company.)*

The *screw drive* is used to drive small cutters (such as the angle milling cutters that are made with a threaded hole). These cutters are screwed onto a *threaded-end arbor* (Figure 14-11). The threaded-end arbor is made with either right-hand or left-hand threads for cutters made with either right-hand or left-hand threads, depending on the desired direction of cutter rotation. Of course, the direction of rotation must be such that cutting torque tends to screw the cutter onto the arbor.

Figure 14-11 An arbor used for angle milling cutters with threaded holes. *(Courtesy Morse Twist Drill & Machine Company.)*

A *fly-cutter arbor* (Figure 14-12) is used to hold a single-tooth cutter with a formed cutting edge. The arbor is slotted to receive the cutter, which is secured in position by setscrews. The cutting edge of the cutter is turned to the desired profile. Clearance is obtained by setting the cutter farther out from the center than the radius to which it was turned.

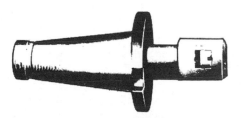

Figure 14-12 Fly-cutter arbor for milling machines with standard spindle end. *(Courtesy Brown & Sharpe Manufacturing Company.)*

Collets

A milling-machine collet is a form of sleeve bushing for reducing the size of the hole in the milling-machine spindle, so that an arbor with a smaller shank can be used (Figure 14-13). Collets are made in several forms, differing in respect to method of drive as follows:

- Tang
- Draw-in
- Clutch

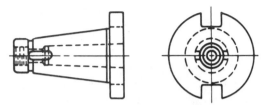

Figure 14-13 Milling-machine collet with draw-in bolt. *(Courtesy Cincinnati Milacon Company.)*

When an arbor is inserted in the collet, the tang projects into the slot. The arbor may be removed by driving a taper key through the slot behind the tang.

A collet holder may be used to take the place of an arbor for holding the cutter. One type of collet holder is inserted in the spindle. It must be removed to change cutters. An extended type of collet holder permits the cutters to be changed without removing the holder from the spindle. A spring-collet holder is used to hold end mills.

Adapters

An adapter is a form of collet that has a standardized spindle end for use on milling machines. A great variety of adapters are available, rendering them suitable for holding various arbors and cutters. Arbor adapters permit the use of arbors, collets, and end mills that have either a Brown & Sharpe taper or a Morse taper on milling machines with the standard spindle end (Figure 14-14).

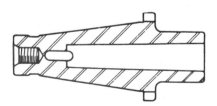

Figure 14-14 Adapter for arbors, collets, and end mills; made with either Brown & Sharpe or Morse-taper shanks for use on milling machines that have a standard spindle end.

(Courtesy Brown & Sharpe Manufacturing Company.)

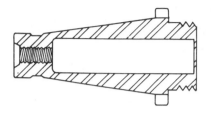

Figure 14-15 Chuck adapter used as a means for attaching chucks on milling machines that have a standard spindle end.

(Courtesy Brown & Sharpe Manufacturing Company.)

A *chuck adapter* (Figure 14-15) has a screw end to provide a means for attaching chucks. This device is used on milling machines that have a standard spindle end.

Several types of adapters are used to hold the various cutters and milling attachment spindles. A standard-taper shank adapter may be used for single-end mills. Still another adapter for end mills that have milling machine standard-taper shanks available with cam-lock is for use with cutter adapters and milling attachment spindles with camlock. These are only a few of the many adapters available for use on milling machines.

Summary

Milling machine operators should be familiar with the holding devices used on the machines that they operate. An inexperienced operator may mistake an arbor (used to hold milling cutters) for a mandrel (used to hold bored parts while turning the outside surface on a lathe).

Three basic devices used on all milling machines are the arbor, collet, and adapter. The arbor is a tapered, cylindrical shaft designed to hold the milling cutters. The holding part of an arbor is cylindrical and clamps the cutter by the use of a nut. The precision and trueness of the arbor influences the accuracy of milling operations.

The collet is a sleeve bushing that reduces the size of the hole in the milling machine spindle so that an arbor with a smaller shank can be used. A collet holder may be used to take the place of an arbor for holding the cutter.

The adapter is a form of collet that has a standardized spindle end for use on the milling machines. Many adapters are available and are styled to hold various arbors and cutters. A standard-taper shank adapter may be used for single-end mills.

Review Questions

1. What is an adapter?
2. What is a collet?
3. How are cutters attached to the arbor?

4. What is an arbor?

5. What is the purpose for bushings on an arbor?

6. Describe the differences between or similarities of the different styles of arbors.

7. What is the difference between friction drive and key drive?

8. What is the arbor sleeve?

9. Why are spacing collars needed on a milling-machine arbor?

10. Describe a chuck adapter and what it does.

Chapter 15

Broaches and Broaching

A *broach* is a straight tool with a series of cutting teeth that gradually increase in size. The broach is used to cut metal and is especially adapted to finishing square, rectangular, or irregularly shaped holes, and for cutting keyways in pulleys, hubs, and so on.

Broaching Principle

Considerable power is required to operate a broach. The broaching operation is used to machine cored or drilled holes to the required shape; one or more broaches are pulled or pushed through these holes. A special tapered cutter, the broach is forced through an opening to enlarge a hole, or alongside a piece of work to shape an exterior. Broaching differs from other machining processes in that a long tool with a series of teeth is used.

A small portion of the metal along the entire cut is removed by each tooth on the broach. The first teeth are smaller, for entering or beginning the cut. The intermediate teeth remove most of the metal, and the last few teeth on the broach finish the hole to proper size (Figure 15-1).

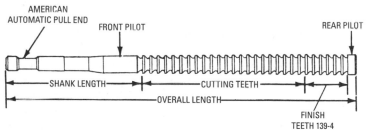

Figure 15-1 Diagram of a typical broach.

A side view of the teeth of a typical broach is shown in Figure 15-2. The *face angle* enables the tooth to cut properly and varies with the material to be cut; it is usually 0° to 20°. The span between the teeth is the *pitch*. The *land* should be of sufficient strength to withstand the cutting strain (may be approximately 25 percent of the pitch). A *straight land* is generally used only at the finishing end of the broach to retain the broach size. The *radius* at

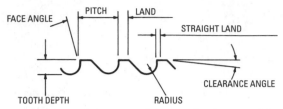

Figure 15-2 Side view of the teeth of a typical broach.

the bottom of the teeth curls the chip and strengthens the broach. The *clearance angle* reduces friction. It should be held to a minimum (about 2°) to prevent excessive wear. The *tooth depth* is proportional to the pitch and should be sufficient to accommodate the chip.

Types of Broaches

A great variety of broaches are available for all types of broaching requirements. Broaching has reduced machining costs and has replaced milling in many situations.

Broaches are commonly made of high-speed steel; carbide-tipped broaches are also used. Molybdenum steel is generally used in broaches.

Broaches may be made either in the solid form or in sections. The sectional broaches may be made in a variety of ways. In some instances, several teeth may be made on a single section and several sections can be used to form the broach.

Shapes of Broaches

Broaches are made in numerous shapes to meet the requirements of all kinds of broaching operations (Figure 15-3). The proportioning of the teeth is important because the cutting operation is progressive. The pitch and other details of tooth design depend on the kind of work for which the broach is intended. In general, two teeth of the broach should be in contact with the work at the same time. Although most broaches have cutting teeth similar to those on milling cutters, a broach may have rounded or smooth teeth. The most common shapes of broaches are as follows:

- Round
- Square
- Keyway (single and double)

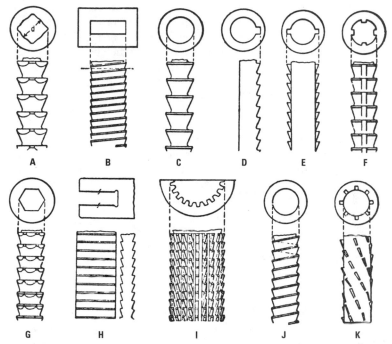

Figure 15-3 The shapes of various types of broaches: (A) square, (B) rectangular, (C) round, (D) single keyway, (E) double keyway, (F) four spline, (G) hexagon, (H) double cut, (I) internal gear, (J) round with helical teeth, (K) helical groove.

- Spline
- Helical

Internal and External Broaches

Internal broaching of round holes is a common commercial broaching operation. Cutting splines in the hubs of gears (or propellers) is commonly done by broaching.

In external (surface) broaching, the broached surface is on the outside of the piece. Broaching takes the place of milling in this type of work. Parts that have irregular external surfaces are well adapted to surface broaching. This type of surface frequently can be finished faster and at less cost by broaching than by milling with a formed milling cutter.

Extreme accuracy can be obtained with a surface broach. Accuracy depends on whether the part is heavy enough to withstand heavy cutting pressures, and on the length of the broach. Accuracy is also greater with a larger number of cutting teeth. The quality of finish is determined by the number of finishing teeth on the broach.

Pull or Push Broaches

Much commercial broaching work is done with the *pull broach* or the *stationary broach*. Pull broaching is used on parts that require a great amount of metal to be removed and on parts that have long finished surfaces. A regular broaching machine is used to pull a long, slender broach through the work. The job is usually completed with one passage of the broach. However, in some instances, two or more broaches of graduated size may be used, especially if considerable stock is to be removed.

Push broaching may be done on an ordinary press by pushing a short, stout broach through the work. Push broaches are shorter and less expensive. This method is used on parts where a small amount of metal is to be removed from the hole and a good finish with close limits is desired.

Care and Sharpening of Broaches

To prevent possible damage to broaches, the teeth should not be allowed to strike a hard object in handling. Minute fractures of the cutting edges of the teeth may result from improper handling. The contour of the cutting teeth should be maintained (that is, original face angle, depth, radius, and straight land should be retained for the tool to cut correctly).

A broach requires sharpening when it no longer provides the finish of which it is capable. Broaches that have a small cross section should be sharpened more often than larger broaches to prevent breakage. When sharpening is necessary, only the front face of the teeth is usually ground. Enough grinding to remove only the slightly rounded edge is necessary. The same amount of stock should be removed from the face of each tooth; thus the step from tooth to tooth is retained.

All sharpening operations should be performed on a rigid machine. Chatter or any tendency of the grinding wheel to vibrate results in poor sharpening. If a tooth should become weakened after numerous sharpenings—or if a portion of a tooth should break for some other reason—several teeth should be retapered to take the added burden, rather than permit the following tooth to assume the additional work.

When internal broaches are used in abrasive metals, considerable land is sometimes formed on the tops of the teeth. Because grinding the face of the teeth does not reduce the land sufficiently, the backoff (clearance) angle of the teeth must be ground. This operation is quite delicate, because the broach must be made to run true between centers. It is recommended that only the roughing teeth should receive this treatment, because any attempt at grinding the finishing teeth in this manner may result in a reduction in size.

Almost all internal broaches are provided with a series of straight teeth to maintain nominal size. Only the first two teeth of this series should be sharpened. The remaining teeth are not sharpened until the previous two teeth have worn; after two teeth have worn, the following two teeth are sharpened, and so on.

When a broach is new, the straight teeth do not have chip breakers. As the broach wears, and the original straight teeth become smaller and are required to remove metal, chip breakers should be added to prevent a wide chip. Wide chips are difficult to remove, and if they are allowed to collect, may cause either tooth breakage or torn holes.

Surface broaches may be resharpened readily by grinding the hook angle or tooth face. Occasionally, they should be ground on the land when they show wear. After resharpening the land, the teeth should again be backed off near the cutting edge of the tooth.

Broaching Machines

A great reduction in velocity must occur between the driving pulley and the broach in order to convert the applied power into the tremendous force necessary to pull or to push a broach in the metal being cut. This great reduction in velocity is obtained by a drive screw and nut (Figure 15-4), and compounded by other gearing.

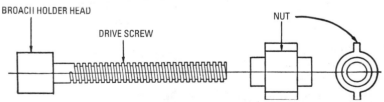

Figure 15-4 Detail of elementary broaching machine illustrating drive screw, broach holder head, and nut.

The drive screw has a broach holder head at one end. A nut is free to turn on the other end of the screw. The screw can move lengthwise only; turning is prevented by suitable guides. The nut is free to turn on the screw, but endwise movement is prevented by stops or thrust bearings placed at the end. Thus, when the nut turns, the drive screw will move one way or the other, depending upon the direction of rotation of the nut. The action is similar to that of the old-fashioned letterpress. This combination gives tremendous leverage, similar to that of a differential hoist. Power is transmitted to turn the nut through a clutch, as shown in detail in Figure 15-5.

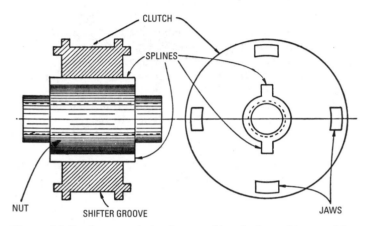

Figure 15-5   A nut and clutch assembly of a broaching machine.

The clutch has jaws on each end and slides on spines so that it may engage either a large or a small gear for both slow drive and quick return. This arrangement is elementary and is not intended to include actual construction details, but only to illustrate principal parts of the elementary machine.

A large gear is on one side of the clutch, and a small gear is on the other side. The large gear meshes with a pinion on the belt-pulley shaft. An intermediate idler gear is interposed between the small return gear and a second gear on the pulley shaft. The intermediate idler gear is provided to reverse the rotation of the nut and the movement of the screw.

In actual operation, the clutch (shown in neutral position in Figure 15-6) is shifted to engage the large drive gear. When power is applied to the tight pulley, it is transmitted to the large drive gear by the pinion and then to the clutch and nut. With this hookup providing first- and second-stage reduction, the pinion makes many revolutions for each revolution of the drive screw. Thus, the high

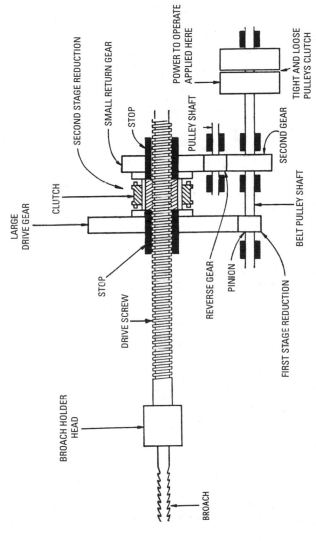

Figure 15-6 Diagram of the essential parts of an elementary broaching machine.

rotative speed and the comparatively weak torque of the pinion are converted into the slow rotative speed and great torque of the nut. The tremendous force necessary to pull or push the broach through the piece being machined is transmitted to the screw. Adjustable tappets and clutch control gear (the means by which the length of stroke may be regulated) are not shown in the diagram.

When the clutch is shifted to engage the small return gear, the rotation of the nut and the movement of the screw are reversed, because of the interposed idler or reverse gear. Moreover, the small size of the return gear gives a quick return. In addition, a hand lever is provided on the machine to start, stop, or reverse the machine by hand control.

Broaching Operations

Broaching has reduced machining costs and has replaced milling in many situations. Originally, broaching was used only for round or irregular holes. It is now used for outside finishing operations, such as outside surfaces of engine cylinder blocks and other pieces used in automobile and aviation engines and parts.

Pull Broaching

In this operation, a part having irregularly shaped holes may be drilled slightly undersized, and a combination broach may be used to finish the inner surfaces. Any desired shape may be machined from this point. Pull broaching is also used to finish parts that have thin walls with a limited amount of resistance to pressure, or on parts that have irregular wall thicknesses. This method is also used extensively for finishing both straight and spiral splines in gears and bushings.

Internal pull broaching can be performed either on horizontal broaching machines or on vertical broaching machines of either the pull-up type or the pull-down type. Semiautomatic and fully automatic broaching machines are desirable for high-production internal broaching operations because they handle the broaches automatically.

Push Broaching

Because it can be performed on an ordinary press, push broaching is a very convenient method of removing only a small amount of metal from a hole. A wide variety of work may be handled in this manner, as a machine may be changed quickly from one type of work to another.

Broaching of round holes in parts where reaming does not give a satisfactory finish, and where the tool cost is too high, is adapted to push broaching. Irregular shapes in die castings and parts made of steel, bronze, Babbitt, or any of the machinable metals may be made with this method. Push broaching is used in sizing the holes in heat-treated gears to remove distortion caused by heat treatment, as well as for small parts that have splines or square, hexagonal, or other irregularly shaped openings.

Surface Broaching

This operation permits greater accuracy and closer limits than other production methods, which results in the production of interchangeable parts that are more easily assembled. Because the surface-broaching tool has an extremely long life, the cost per piece is greatly reduced in comparison to other production methods. The longer life is due to the substantial support back of the surface-broaching tool, the elimination of vibration, the slow cutting speed, and its positive cutting action.

A shear angle is used to facilitate the cutting action wherever possible. Chip breakers are essential on the roughing teeth for breaking the heavy cut and for making narrow chips that will readily fall out of the broach teeth on completion of the cut.

Several types of surface broaches are available for unusual types of surface-broaching jobs. Surface broaches are made in plain slab or special types for many irregularly shaped and formed cuts that cannot be machined by any other method. These broaches are usually made in sections and then placed in a substantial holder. The holder is attached to a subholder or to the machine slide itself.

The surface broach has the cutting quality of both a roughing tool and a finishing tool built into a single cutting unit. The roughing section of the broach removes the major portion of the metal. The teeth are evenly graduated for size (or amount of metal to be removed) from the roughing section to the finishing teeth, which are reserved for light-duty cutting, for accurate sizing of the work, and for producing a fine finish at the same time. The work has an opportunity to cool slightly as it passes from the roughing teeth to the finishing teeth. Therefore, any error due to temperature change or springing of the work is eliminated.

Round Broaching

The chief advantages of round broaching are speed, uniformity of size of holes, and long tool life. When broaching round holes and other shapes, each tooth of the broach passes the work only once. The chip rarely exceeds 0.003 inch. This explains the long tool life of the broach.

If an extremely fine finish is desired, a burnishing section can be added to the broach. Pilot sections placed between the broach cutting teeth make it possible to finish the broached hole concentric with the starting hole. These pilots permit the removal of a small amount of stock, removing it from all sides, as in sizing operations. Of course, the smaller the amount of stock there is to be removed from the hole, the finer should be the finish before broaching. Likewise, the broach may be shorter and cheaper for the smaller amounts of stock to be removed.

Summary

A broach is a straight tool with a series of teeth that gradually increase in size. The broach is used to cut metal, especially for square, rectangular, and irregularly shaped holes. Considerable power is needed to operate a broach.

Broaches are made in various shapes to meet the requirements of all kinds of operations. The most common shapes of broaches are round, square, keyway, spline, and helical. Broaches may be made either in the solid form or in sections. In some cases, several teeth may be made on a single section and several sections can be used to form the broach.

Great reduction in velocity must occur between the driving pulley and the broach to convert the applied power into the tremendous force necessary to pull or push the broach through the metal being cut. This force is obtained by a drive screw and nut, and is compounded by other gearing.

Considerable power is required to operate a broach. The broaching operation is used to machine cored or drilled holes to the required shape; one or more broaches are pulled or pushed through these holes. The broach, a special tapered cutter, is forced through an opening to enlarge a hole, or alongside a piece of work to shape an exterior. Broaching differs from other machining processes in that a long tool with a series of teeth is used.

Broaching has reduced machining costs and has replaced milling in many situations. It was used only for round or irregular holes. Now, it is used for outside finishing operations (such as outside surfaces of engine cylinder blocks and other auto parts).

Review Questions

1. What is broaching?
2. Name the most common shapes of broaches.
3. How are broaches made?

4. Explain why broaching has cut operation cost in machine shops.

5. What is the face angle in a broach?

6. Define the term *straight land*.

7. What is meant by tooth depth?

8. Broaches may be made in solid form or in _____.

9. List five shapes for broaches.

10. Much commercial broaching work is done with the _____ broach or the stationary broach.

11. Surface broaches may be resharpened readily by grinding the hook angle or _____ face.

12. Describe pull broaching.

13. Describe push broaching.

14. What is surface broaching?

15. What is the chief advantage of round broaching?

Chapter 16

Electrical Safety in the Machine Shop

The machine shop usually has a number of machines that require electrical circuitry and control devices that may, under certain conditions, cause damage to property and human operators. It is best to be prepared by knowing something about electrical requirements of machines and what should be done if the machines malfunction.

Power Sources

Commercial alternating current (AC) is usually utilized by the machine shop since it would be somewhat expensive to generate the electricity locally with gasoline- or diesel-engine generators. Commercial AC may be generated by falling-water generators, nuclear-powered generators, or fossil-fuel–powered generators. In Alaska, however, where long distribution lines are impractical because of ice and wind conditions that result in line damage, engine-driven generators are common.

Once electricity is generated in sufficient amounts for consumption in large quantities, the second necessary step is to get the energy to the consumer. Once the wires are strung and the equipment for distribution is installed, the lines usually end at the inside of the machine shop distribution panel (Figure 16-1).

Three-Phase Power

Most larger machine shops use three-phase power, which means that it takes three wires. Three-phase power can be generated by a wye-connected generator or a delta-connected generator. Instead of having six lines coming from the three-phase generator, one of the leads from each phase can be connected to form a common junction. The stator is then called a *wye*. Sometimes the wye connection is also called a *star connection*. The common lead may not be brought out of the generator. If it is brought out, it is called the *neutral* (Figure 16-2). If the wye connection has the neutral brought out of the generator, it is called a *four-wire, three-phase system*.

The wye connection provides 1.73 times the phase voltage for any two of the three wires connected. The line currents are equal to

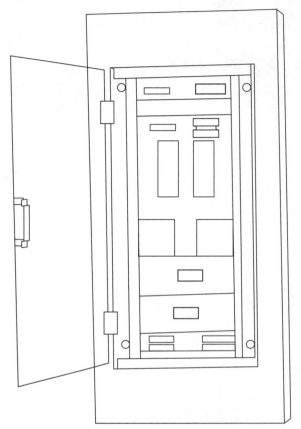

Figure 16-1 Electrical distribution panel.

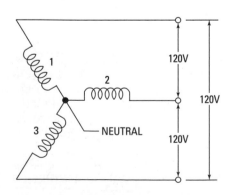

Figure 16-2 Three-phase power wye or star connections.

the current in any phase. The advantage of a wye connection is its ability to produce more voltage. Note in the delta connections in Figure 16-3 that windings 1 and 2 are in series with each other. Windings 2 and 3 are also in series with each other. If windings 1 and 2 are used for connections, they, too, are in series. Thus, no matter which connections are used, there are two coils in series to produce the single-phase power needed from a three-phase line.

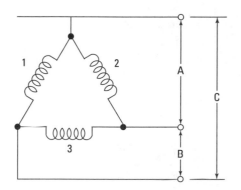

Figure 16-3 Three-phase power delta connections.

The delta connection provides 1.73 times the phase current for any two of the three wires connected.

Each connection method may be used, so it is best to know which you are dealing with before becoming aware of some of the problems associated with the motors that drive the equipment.

Power Panels

Panelboards are available from many different manufacturers. A wide range of panel types are available for applications up to 600 volts AC with 10,000 through 200,000 amperes maximum short-circuit current rating. Plug-in or bolt-on branch circuit breakers are available. Plug-in circuits are locked in position with a dead front-panel cover, assuring positive contact. Figure 16-4 shows what the typical panel wiring diagram looks like. The numbers on the side represent the connection points for various circuits.

Raceways and Cable Trays

For the safety of persons working in an area with high voltages and high currents (as is often required by machine shop equipment), it is best to enclose the wiring in raceways or place the wires on cable trays in an organized and safe manner. An economical raceway system designed for support and protection of electrical wire and cable

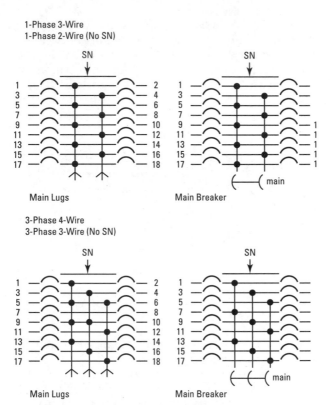

Figure 16-4 Panel wiring diagrams.

is available in aluminum and two types of galvanized steel for outdoor or indoor applications (Figure 16-5). Examples of how the trays can be routed and used to support heavy cable are shown. The trays are made in straight sections, with matching fittings to accommodate all changes of direction or quantity of cables. They are usually made of aluminum or zinc-coated steel.

Trough-type trays protect cables from damage and give good support and ample ventilation. Solid-bottom fittings generally create no ventilation problems, since they are a small part of the system. Cables are adequately ventilated through straight sections. Ladder trays provide maximum ventilation to power cables and other heat-producing cables.

However, cables are vulnerable to damage and covers are available. Various parts are needed to support the trays and covers.

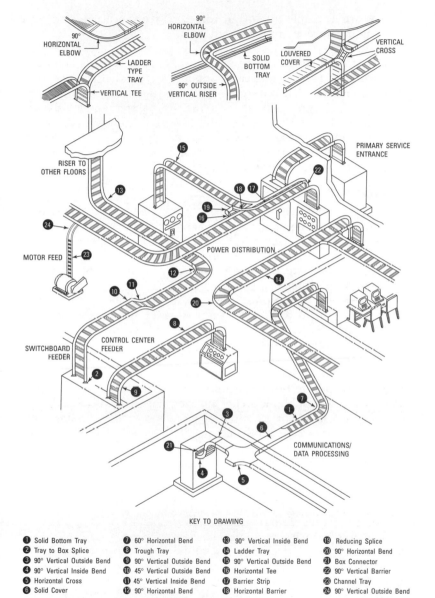

Figure 16-5 Cable tray configuration.

KEY TO DRAWING

❶ Solid Bottom Tray	❼ 60° Horizontal Bend
❷ Tray to Box Splice	❽ Trough Tray
❸ 90° Vertical Outside Bend	❾ 90° Vertical Outside Bend
❹ 90° Vertical Inside Bend	❿ 45° Vertical Outside Bend
❺ Horizontal Cross	⓫ 45° Vertical Inside Bend
❻ Solid Cover	⓬ 90° Horizontal Bend

⓭ 90° Vertical Inside Bend	⓳ Reducing Splice
⓮ Ladder Tray	⓴ 90° Horizontal Bend
⓯ 90° Vertical Outside Bend	㉑ Box Connector
⓰ Horizontal Tee	㉒ 90° Vertical Barrier
⓱ Barrier Strip	㉓ Channel Tray
⓲ Horizontal Barrier	㉔ 90° Vertical Outside Bend

Cables are available for use in cable trays marked CT (for cable tray) on the outside of the jacket.

The cable system must be complete. It must be used as a complete system of straight sections, angles, offsets, saddles, and other associated parts to form a cable support system that is continuous and grounded as required by the *National Electrical Code*. The system must be grounded as any raceway system must also be grounded. The Code treats the cable tray as a raceway and a wiring method. Limitations are placed on the number, size, and placement of conductors inside the tray. These limitations can be obtained by checking the *National Electrical Code*.

Motor Problems

One of the most frequent problems associated with working on a piece of electrical equipment in a machine shop is the motor used to power or drive the machines. Some common problems may be easily identified by the machinist and pointed out to the electrician. This could eliminate any unsafe conditions and possibly speed up the correction problem.

Easy-to-detect symptoms, in many cases, indicate exactly what is wrong with fractional-horsepower motors. However, where general types of trouble have similar symptoms, it becomes necessary to check each possible cause separately. Table 16-1 lists some of the more common ailments of small motors, together with suggestions as to possible causes. Most common motor problems can be checked by some test or inspection. The order of making these tests rests with the troubleshooter. However, it is natural to make the simplest test first. For example, when a motor fails to start, you would first inspect the motor connections, since this is an easy and simple thing to do.

Table 16-1   Squirrel Cage Motor Problems

Symptom and Possible Cause	Possible Remedy
Motor Will Not Start	
(a) Overload control tripped	(a) Wait for overload to cool. Try starting again. If motor still does not start, check all the causes as outlined in the following.
(b) Power not connected	(b) Connect power to control and control to motor. Check clip contacts.
(c) Faulty (open) fuses	(c) Test fuses.

Table 16-1 *(continued)*

Symptom and Possible Cause	Possible Remedy
(d) Low voltage	(d) Check motor nameplate values with power supply. Also check voltage at motor terminals with motor under load to be sure wire size is adequate.
(e) Wrong control connections	(e) Check connections with control wiring diagram.
(f) Loose terminal lead connection	(f) Tighten connections.
(g) Driven machine locked	(g) Disconnect motor from load. If motor starts satisfactorily, check driven machine.
(h) Open circuit in stator or rotor winding	(h) Check for open circuits.
(i) Short circuit in stator winding	(i) Check for shorted coil.
(j) Winding grounded	(j) Test for grounded winding.
(k) Bearings stiff	(k) Free bearings or replace.
(l) Grease too stiff	(l) Use special lubricant for special conditions.
(m) Faulty control	(m) Check control wiring.
(n) Overload	(n) Reduce load.
Motor Noisy	
(a) Motor running single-phase	(a) Stop motor, and then try to start. (It will not start on single-phase.) Check for "open" in one of the lines or circuits.
(b) Electrical load imbalanced	(b) Check current balance.
(c) Shaft bumping (sleeve-bearing motors)	(c) Check alignment and condition of belt. On pedestal-mounted bearing, check end play and axial centering of rotor.
(d) Vibration	(d) Driven machine may be imbalanced. Remove motor from load. If motor is still noisy, rebalance rotor.
(e) Air gap not uniform	(e) Center the rotor and, if necessary, replace bearings.
(f) Noisy ball bearings	(f) Check lubrication. Replace bearings if noise is persistent and excessive.

(continued)

Table 16-1 *(continued)*

Symptom and Possible Cause	Possible Remedy
(g) Loose punchings or loose rotor on shaft	(g) Tighten all holding bolts.
(h) Rotor rubbing on stator	(h) Center the rotor and replace bearings if necessary.
(i) Object caught between fan and end shields	(i) Disassemble motor and clean. Any rubbish around motor should be removed.
(j) Motor loose on foundation	(j) Tighten hold-down bolts. Motor may possibly have to be realigned.
(k) Coupling loose	(k) Insert feelers at four places in coupling joint before pulling up bolts to check alignment. Tighten coupling bolts securely.

Motor Running Temperature Too High

(a) Overload	(a) Measure motor loading with ammeter. Reduce load.
(b) Electrical load imbalanced (fuse blown, faulty control, etc.)	(b) Check for voltage imbalance or single phasing. Check for "open" in one of the lines or circuits.
(c) Restricted ventilation	(c) Clean air passages and windings.
(d) Incorrect voltage and frequency	(d) Check motor nameplate values with power supply. Also check voltage at motor terminals with motor under full load.
(e) Motor stalled by driven machine or by tight bearings	(e) Remove power from motor. Check machine for cause of stalling.
(f) Stator winding shorted or grounded	(f) Test winding for short circuit or ground.
(g) Rotor winding with loose connections	(g) Tighten, if possible, or replace with another rotor.
(h) Motor used for rapid reversing service	(h) Replace with motor designed for this service.
(i) Belt too tight	(i) Remove excessive pressure on bearings.

Bearings Hot

(a) End shields loose or not replaced properly	(a) Make sure end shields fit squarely and are properly tightened.

Table 16-1 *(continued)*

Symptom and Possible Cause	Possible Remedy
(b) Excessive belt tension or excessive gear slide thrust	(b) Reduce belt tension or gear pressure and realign shafts. See that thrust is not being transferred to motor bearings.
(c) Bent shaft	(c) Straighten shaft or replace.
Sleeve Bearings Hot	
(a) Insufficient oil	(a) Add oil. If oil supply is very low, drain, flush, and refill.
(b) Foreign material in oil, or poor grade of oil	(b) Drain oil, flush, and relubricate, using industrial lubricant recommended by a reliable oil company.
(c) Oil rings rotating slowly or not rotating at all	(c) Oil too heavy; drain and replace.
(d) Motor tilted too far	(d) Level motor or reduce tilt and realign, if necessary.
(e) Oil rings bent or otherwise damaged in reassembling	(e) Replace oil rings.
(f) Oil ring out of slot	(f) Adjust or replace retaining clip.
(g) Motor tilted, causing end thrust	(g) Level motor, reduce thrust, or use motor designed for thrust.
(h) Defective bearings or rough shaft	(h) Replace bearings. Resurface shaft.
Ball Bearings Hot	
(a) Too much grease	(a) Remove relief plug and let motor run. If excess grease does not come out, flush and relubricate.
(b) Wrong grade of grease	(b) Add proper grease.
(c) Insufficient grease	(c) Remove relief plug and regrease bearing.
(d) Foreign material in grease	(d) Flush bearings, relubricate; make sure that grease supply is clean. Keep can covered when not in use.
(e) Bearings misaligned	(e) Align motor and check bearings housing assembly. See that the bearings races are exactly 90° with shaft.
(f) Bearings damaged	(f) Replace bearings.

A combination of systems will often give a definite clue to the source of the trouble and hence eliminate other possibilities. For example, in the case just cited where a motor won't start if heating occurs, it offers the suggestion that a short or ground exists in one of the windings and eliminates the likelihood of an open circuit, poor line connection, or defective starter switch.

Centrifugal starting switches, found in many types of single-phase fractional-horsepower motors, occasionally are a source of trouble (Figure 16-6). If the mechanism sticks in the run position, the motor will not start. On the other hand, if stuck in the closed position, the motor will not attain speed and the start winding heats up rapidly. The motor may also fail to start if the contact points of the switch are out of adjustment or coated with oxide. It is important to remember, however, that any adjustment of the switch or contacts should be made only at the factory or by authorized service personnel.

Three-Phase Motor Symptoms

If one leg of the three-phase power opens while the motor is running, it will slow down and hum, never coming up to speed. This may indicate that the fuse in that leg has opened. If the motor does not start and hums, it is possible to start the motor physically by turning the motor shaft or pulley attached to it. However, this means the motor will not come up to speed and will hum excessively. It is also possible to start the motor in clockwise or counterclockwise direction if one leg of the power source is open. Continued use at slow speed can cause overheating of the motor.

The *National Electrical Code* has established standards that apply to the control and protection of motors and their associated circuits. For example, Figure 16-7 is a general circuit for the installation of squirrel cage induction motors. See Table 16-1 for problems with squirrel cage motors.

Motor controllers and their associated circuits will vary among different models as well as different controller manufacturers. Figure 16-8 is a functional schematic diagram of a motor controller that incorporates overload protection. Good engineering practice emphasizes using three overload relays (one in each phase line of the motor) to provide protection from voltage imbalance conditions. The size of the controller and overload protection relays must be in accord with the controller manufacturer's specifications as well as the *National Electrical Code*.

As a further safeguard against motor failure caused by prolonged locked rotor conditions, the overload protection should trip out in 15 seconds or less at lock-rotor current.

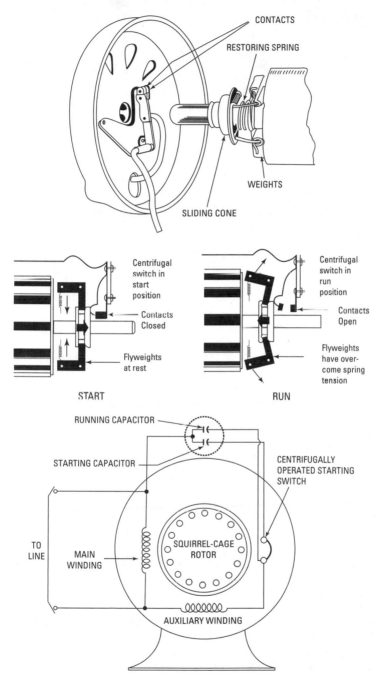

Figure 16-6 Centrifugal switch on a single-phase motor.

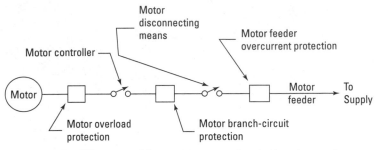

Figure 16-7 Diagram with motor protective devices located.

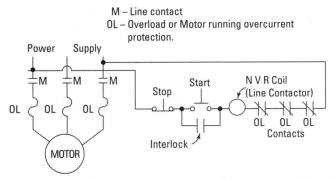

Figure 16-8 Motor control consists of a contactor (M) and overload relays (OL).

DC Motors

One of the problems associated with DC motors is high maintenance costs in both time and materials. They do have a tendency to require attention at the commutator and brush level. Figure 16-9 shows some of the problems associated with worn or damaged commutators. Brush sparking can cause situations that may lead to explosions in certain atmospheres.

More maintenance is required by motors with commutators. High-speed, series-wound motors should not be used on long, continuous-duty cycle applications, because the commutator and brushes can potentially cause trouble. Gummy commutators and oil-soaked brushes can cause sluggish action and severe sparking. The commutator can be cleaned with fine sandpaper. However, if pitted spots still appear, the commutator should be reground.

(A)

PITCH BAR-MARKING produces low or burned spots on the commutator surface that equals half or all the number of poles on the motor.

(B)

STREAKING on the commutator surface denotes the beginning of serious metal transfer to the carbon brush.

(C)

HEAVY SLOT BAR-MARKING involves etching of the trailing edge of commutator bar in relation to the numbered conductors per slot.

(D)

THREADING of commutator with fine lines is a result of excessive metal transfer leading to resurfacing and excessive brush wear.

(E)

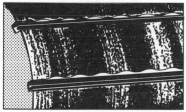

COPPER DRAG is an abnormal amount of excessive commutator material at the trailing edge of bar. Even though rare, flashover may occur if not corrected.

(F)

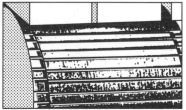

GROOVING is caused by an abrasive material in the brush or atmosphere.

Figure 16-9 Worn or damaged commutators.

Motor Lubrication

Overheated motors can be caused by lack of proper lubrication. The lubricant used with sleeve bearings must actually provide an oil film that completely separates the bearing surface from the rotating shaft member and, ideally, eliminates metal-to-metal contact.

The selection of the oil that will provide the most effective bearing lubrication and not require frequent renewal merits careful consideration. Good lubricants are essential to low maintenance costs. Top-grade oils are recommended because they are refined from pure petroleum; they are substantially noncorrosive as far as metal surfaces to be lubricated are concerned; they are free from sediment, dirt, or other foreign materials; and they are stable with respect to heat and moisture encountered in the motor. In performance terms, the higher-priced oils prove to be cheaper in the long run.

Sleeve bearings are less sensitive to a limited amount of abrasive or foreign materials than are ball bearings. This is because of the ability of the relatively soft surface of the sleeve bearing to absorb hard particles of foreign materials. Good maintenance practice recommends that the oil and bearing be kept clean. Frequency of oil changing will depend on local conditions such as severity and continuity of service and operating temperature. A conservative lubrication maintenance program should call for periodic inspections of the oil level and cleaning and refilling with new oil every six months.

WARNING
Over-lubrication should be avoided. Insulation damage by excessive motor lubricant represents one of the most common causes of motor-winding insulation failure. This is true for both sleeve and ball-bearing motors.

Drum Switches

One of the most commonly used switches for motor control is the drum switch (Figure 16-10). It has the capacity to reverse a motor or turn it off. Drum switches may be used to control the starting and reversing of AC polyphase, AC single-phase, or DC motors directly across the line. They are compact and inexpensive, but ruggedly constructed.

Incorrect use of switching can cause overheating and motor damage. By quickly turning the drum switch first in one direction and then the other, it is possible to cause motor overheating if done repeatedly for any period of time.

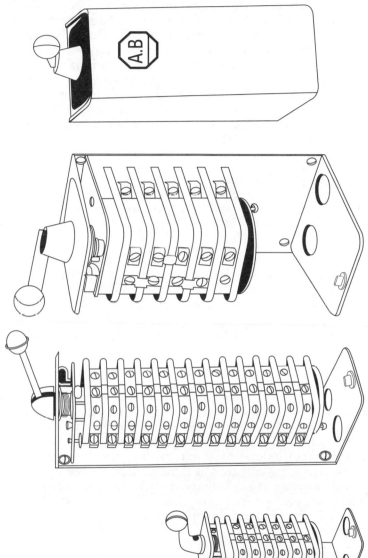

Figure 16-10 Drum switches—covered and uncovered.

Safety in the Shop

Electrical shock can be deadly. It is not the voltage, but the current that kills. People have been killed by 110 volts in the home and with as little as 42 volts of direct current (DC). The real measure of a shock's intensity lies in the amount of current forced through the body. Any electrical device used on a house wiring circuit can, under certain conditions, transmit a fatal current. Since you do not know how much current went through the body in an electrical accident, it is necessary to perform artificial respiration to try to get the person breathing again, or if the heart is not beating, cardiopulmonary resuscitation (CPR).

NOTE

A heart that is in fibrillation cannot be restricted by closed-chest cardiac massage. A special device called a defibrillator is available in some medical facilities and by ambulance services, as well as in some high schools where at least one teacher is skilled in the use of the machine.

Muscular contractions are so severe with 200 mA and above that the heart is forcibly clamped during the shock. Keep in mind that 200 mA is 0.2 amperes. Clamping prevents the heart from going into ventricular fibrillation, making the victim's chances for survival better.

Working with or around electricity can be dangerous. However, electricity can be safe if properly respected.

Fire Extinguishers

Fire extinguishers are limited in their application. They are used to control small fires that are identified properly by the person selecting the extinguisher. Extinguishers are made for various uses, as shown in Table 16-2. Needless to say, the water-type extinguisher is not useful for electrical fires. Electrical fires are classified as Class C. Class C fires are described as those dealing with energized electrical equipment; that is, where the electrical nonconductivity of the extinguishing medium is of importance. Electrical fires call for carbon dioxide, dry chemicals, multipurpose dry chemical, and Halon 1211 types of extinguisher.

Summary

The machine shop usually has a number of machines that require electrical circuitry and control devices that may under certain conditions cause damage to property and human operators. It is best to be prepared by knowing something about electrical requirements of machines and what should be done if they malfunction.

Most larger machine shops use three-phase power. That means it takes three wires. The wye connection provides 1.73 times the phase voltage for any two of the three wires connected. The line currents are equal to the current in any phase. The advantage of a wye connection is its ability to produce more voltage. The delta connection provides 1.73 times the phase current for any two of the three wires connected.

Each connection method may be used, so it is best to know which you are dealing with before becoming aware of some of the problems associated with the motors that drive the equipment.

For the safety of persons working in an area with high voltages and high currents (as is often required by machine shop equipment), it is best to enclose the wiring in raceways or place the wires on cable trays in an organized and safe manner. Cables are vulnerable to damage and covers are available. Various parts are needed to support the trays and covers. Cables are available for use in cable trays marked CT (for cable tray) on the outside of the jacket.

One of the most frequent problems associated with working in a machine shop on a piece of electrical equipment is the motor used to power or drive the machines. Some common problems may be easily identified by the machinist and pointed out to the electrician and thereby eliminate any unsafe conditions and possibly speed up the correction problem.

Easy-to-detect symptoms, in many cases, indicate exactly what is wrong with fractional-horsepower motors. However, where general types of trouble have similar symptoms, it becomes necessary to check each possible cause separately. If one leg of the three-phase power opens while the motor is running, it will slow down and hum, never coming up to speed. This may indicate that the fuse in that leg has opened. If the motor does not start and hums, it is possible to start the motor physically by turning the motor shaft or pulley attached to it. However, this means the motor will not come up to speed and will hum excessively. It is also possible to start the motor in clockwise or counterclockwise direction if one leg of the power source is open. Continued use at slow speed can cause overheating of the motor.

One of the problems associated with DC motors is high maintenance costs in both time and materials. They do have a tendency to need attention at the commutator and brush level. Many of the overheated motors occur because of lack of proper lubrication. The lubricant used with sleeve bearings must actually provide an oil

Table 16-2 Fire Extinguishers

Extinguisher Classifications†	A — Water Types (includes antifreeze)		AB — AFFF Foam and FFFP	BC — Carbon Dioxide	BC — Dry Chemical Types Purple K / Super K / Monnex / Potassium Bicarb Urea based		BC — Halogenated Types 1211 1301 1211/1301	ABC — Multipurpose Dry Chemical		ABC — Halogenated Types 1211 1211/1301	D — Dry Powder
Discharge Method	Stored Pressure	Pump Tank	Stored Pressure	Self Expelling	Stored Pressure	Cartridge Operated	Stored Pressure	Stored Pressure	Cartridge Operated	Stored Pressure	Cartridge Operated
Sizes Available	2½ gal.	2½–5 gal.	2½ gal. (33 gal.)	5–20 lb (50–100 lb)	2½–30 lb (50–350 lb)	4–30 lb (125–350 lb)	1–5 lb	2½–20 lb (50–350 lb)	5–30 lb (125–350 lb)	5½–22 lb (50–150 lb)	30 lb (150–350 lb)
Horizontal Range (Approx.)	30–40 ft	30–40 ft	10–25 ft (30 ft)	3–8 ft (3–10 ft)	10–15 ft (15–45 ft)	10–20 ft (15–45 ft)	10–16 ft	10–15 ft (15–45 ft)	10–20 ft (15–45 ft)	9–16 ft (20–35 ft)	5 ft (15 ft)
Discharge Time (Approx.)	1 min.	1–3 min.	50–65 sec. (1 min.)	8–15 sec. (10–30 sec.)	8–25 sec. (25–60 sec.)	8–25 sec. (25–60 sec.)		8–25 sec. (20–60 sec.)	8–25 sec. (25–60 sec.)	10–18 sec. (30–45 sec.)	20 sec. (150 lb, 70 sec. 350 lb, 1¾ min.)
Operating Precautions and Agent Limitations	Conductor of electricity. Needs protection from freezing, (except antifreeze). Use on flamable liquids and grease will spread fire.		Conductor of electricity. Needs protection from freezing. Not	Smothering occurs in high concentrations. Avoid contact	Extensive cleanup, particularly on delicate electronic equipment. Obscures visibility in confined spaces.		Avoid high concentrations and unnecessary use.	Extensive cleanup. Damages electronic equipment. Obscures visibility in confined spaces. Limited penetrating ability on deep-seated Class A fires.		Avoid high concentrations and unnecessary use.	Not listed.

effective on water-soluble flammable liquids such as alcohol, unless otherwise stated on nameplate. AFFF not effective on pressurized flammable liquid/gas fires.

with discharge horn. Limited effectiveness under windy conditions. Severely reduced effectiveness at sub-zero (F) temperatures.

†NOTE: Only dry chemical types are effective on pressurized flammable gases and liquids; for deep fat fryers, multipurpose ABC dry chemicals are not acceptable.

NOTE: Protection required below 40°F and above 120°F.

NOTE: These photos are not proportional in relation to one another.

A AB CO_2 BC ABC D

Courtesy National Association of Fire Equipment Distributors, Chicago, IL.

film that completely separates the bearing surface from the rotating shaft member and, ideally, eliminates metal-to-metal contact.

The selection of the oil that will provide the most effective bearing lubrication and not require frequent renewal merits careful consideration. Good lubricants are essential to low maintenance costs. Sleeve bearings are less sensitive to a limited amount of abrasive or foreign materials than are ball bearings. This is because of the ability of the relatively soft surface of the sleeve bearing to absorb hard particles of foreign materials.

One of the most commonly used switches for motor control is the drum switch. It has the capacity to reverse a motor or turn it off. Drum switches may be used to control the starting and reversing of AC polyphase, AC single-phase, or DC motors directly across the line. They are compact and inexpensive, but ruggedly constructed.

Electrical shock can be deadly. It is not the voltage, but the current that kills. A heart that is in fibrillation cannot be restricted by closed-chest cardiac massage. A special device called a defibrillator is available in some medical facilities and schools. Working with or around electricity can be dangerous. However, electricity can be safe if properly respected.

Fire extinguishers are limited in their application. They are used to control small fires that are identified properly by the person selecting the extinguisher. Electrical fires call for carbon dioxide, dry chemicals, multipurpose dry chemical, and Halon 1211 types of extinguishers.

Review Questions

1. How is commercial electrical power generated?
2. What is three-phase power?
3. Where are the terms *delta* and *wye* used?
4. What is the advantage of raceways in a machine shop?
5. What is the difference between a raceway and a cable tray?
6. If one leg of the three-phase power is out, what is the symptom noticed in the electric motor?
7. What is the main problem associated with DC motors?
8. Why is proper motor lubrication important?
9. Why should over-lubrication be avoided?
10. What is one of the most commonly used switches in electric motor control?

Appendix

Reference Materials

When dealing with conversions to and from the metric system, keep the following in mind:

- The metric unit of length is the meter, which is equivalent to 39.37 inches.
- The metric unit of weight is the gram, which is equivalent to 15.432 grains.
- The following prefixes are used for subdivisions and multiples:

  milli—$\frac{1}{1000}$

  centi—$\frac{1}{100}$

  deci—$\frac{1}{10}$

  deca—10

  hecto—100

  kilo—1000

  myria—10,000

This appendix provides a reference for the following:

- Miscellaneous useful facts
- Metric conversions
- Water factor conversions
- Weights of various sizes of steel bars

Miscellaneous Useful Facts

Following are some useful facts to keep in mind:

- *Area of a circle*—Multiply half the circumference of a circle by half its diameter.
- *Area of a circle*—Multiply the square of the circumference of a circle by 0.07958.
- *Area of a circle*—Multiply the square of the diameter by 0.7854.

- *Area of the surface of a ball (sphere)*—Multiply the square of the diameter by 3.1416.
- *Circumference of a circle*—Multiply the diameter by 3.1416.
- *Circumference of a circle*—Multiply the radius by 6.283185.
- *Diameter of a circle equal in area to a given square*—Multiply a side of the square by 1.12838.
- *Diameter of a circle inscribed in a hexagon*—Multiply a side of the hexagon by 1.7321.
- *Diameter of a circle inscribed in an equilateral triangle*—Multiply a side of the triangle by 0.57735.
- *Diameter of a circle*—Multiply the circumference by 0.31831.
- *Diameter of a circle*—Multiply the square root of the area of a circle by 1.12838.
- *Radius of a circle*—Multiply the circumference by 0.159155.
- *Radius of a circle*—Multiply the square root of the area of a circle by 0.56419.
- *Side of a hexagon inscribed in a circle*—Multiply the diameter of the circle by 0.500.
- *Side of a square equal in area to a given circle*—Multiply the diameter by 0.8862.
- *Side of a square inscribed in a circle*—Multiply the diameter by 0.7071.
- *Side of an equilateral triangle inscribed in a circle*—Multiply the diameter of the circle by 0.866.
- *Volume of a ball (sphere)*—Multiply the cube of the diameter by 0.5236.

Metric Conversions

This section provides useful information about converting to and from the metric measurement system, including the following:

- Metric and English equivalent measures (including measures of length, weight, and capacity)
- English conversions (including length, area, volume, weight, energy, pressure, and power)
- Metric conversions (including length, area, volume, weight, unit weight, pressure, energy, power, and miscellaneous)

Metric and English Equivalent Measures

This section provides a basis to convert from metric to English measures.

Length

1 meter = 39.37 inches or 3.28083 feet or 1.09361 yards

0.3048 meter = 1 foot

1 centimeter = 0.3937 inch

2.54 centimeters = 1 inch

1 millimeter = 0.03937 inch or nearly $\frac{1}{25}$ inch

25.4 millimeters = 1 inch

1 kilometer = 1093.61 yards, or 0.62137 mile

NOTE

The ratio 25.4 mm = 1 inch is used to convert millimeters to inches.

Weight

1 gram = 15.432 grains

0.0648 gram = 1 grain

28.35 grams = 1 ounce avoirdupois

1 kilogram = 2.2046 pounds

0.4536 kilogram = 1 pound

1 metric ton (or 1000 kilograms) = 0.9842 ton of 2240 pounds (long tons) or 19.68 cwt. or 2204.6 pounds

1.016 metric tons (1016 kilograms) = 1 ton of 2240 pounds long ton

Capacity

1 liter (or 1 cubic decimeter) = 61.023 cubic inches or 0.03531 cubic foot or 0.2642 gal. (American) or 2.202 lb of water at 62°F

28.317 liters = 1 cubic foot

3.785 liters = 1 gallon (American)

4.543 liters = 1 gallon (Imperial)

English Conversion Table
This section provides a guide for converting English measures.

Length
Inches = Yards × 36.00

Inches = Miles × 63360.00

Feet = Inches × 0.0833

Feet = Yards × 3.00

Feet = Miles × 5280.00

Yard = Inches × 0.02778

Yards = Feet × 0.3333

Yards = Miles × 1760.00

Miles = Inches × 0.00001578

Miles = Feet × 0.0001894

Miles = Yards × 0.0005681

Circumference = Diameter of circle × 3.1416

Diameter = Circumference of circle × 0.3188

Area
Square inches = Square feet × 144.00

Square inches = Square yards × 1296.00

Square feet = Square inches × 0.00694

Square feet = Square yards × 9.00

Square yards = Square inches × 0.0007716

Square yards = Square feet × 0.11111

Area = Diameter of circle squared × 0.7854

Surface = Diameter of sphere squared × 3.1416

Volume
Cubic inches = Cubic feet × 1728.00

Cubic feet = Cubic inches × 0.0005787

Cubic feet = Cubic yards × 27.00

Cubic yards = Cubic inches × 0.00002143

Cubic yards = Cubic feet × 0.03704

U.S. gallons = Cubic inches × 0.004329

U.S. gallons = Cubic feet × 7.4805
Volume = Diameter of sphere cubed × 0.5236

Weight

Ounces = Grains (avoirdupois) × 0.002286
Ounces = Pounds (avoirdupois) × 16.00
Ounces = Tons (avoirdupois) × 32000.00
Pounds = Ounces (avoirdupois) × 0.0625
Pounds = Tons (avoirdupois) × 2000.00
Hundredweight = Pounds (avoirdupois) × 0.01
Tons = Ounces (avoirdupois) × 0.00003125
Tons = Pounds (avoirdupois) × 0.0005

Energy

Btu per minute = Ton of refrigeration × 200.00
Feet pounds = Btu × 778.26
Feet pounds per minute = Horsepower × 33000.00

Pressure

Inches of water (60°F) = Pounds per square inch × 27.70
Inches of Mercury (60°F) = Pounds per square inch × 2.041
Feet of water (60°F) = Pounds per square inch × 2.31
Pounds per square inch = Feet of water (60°F.) × 0.433
Pounds per square inch = Inches of Mercury (60°F) × 0.490
Pounds per square inch = Inches of water (60°F) × 0.0361

Power

Watts = Horsepower × 746
Horsepower = Watts × 0.001341
Btu per minute = Horsepower × 42.4

Standard Metric to English Conversions

This section provides a guide for converting metric measurements to English measurements.

Length

Inches = Millimeters × 0.03937
Inches = Millimeters ÷ 25.4
Inches = Centimeters × 0.3937
Inches = Centimeters ÷ 2.54
Inches = Meters × 39.37
Feet = Meters × 3.281
Feet = Kilometers × 3280.8
Yards = Meters × 1.0936
Miles = Kilometers × 0.6214
Miles = Kilometers ÷ 1.6093

Area

Square inches = Square millimeters × 0.00155
Square inches = Square millimeters ÷ 645.2
Square inches = Square centimeters × 0.155
Square inches = Square centimeters ÷ 6.452
Square feet = Square meters × 10.764
Acres = Square kilometers × 247.1
Acres = Hectares × 2.471

Volume

Cubic inches = Cubic centimeters ÷ 16.387
Cubic inches = Liters × 61.023
Cubic feet = Cubic meters × 35.314
Cubic feet = Hectoliters × 3.531
Cubic feet = Liters ÷ 28.317
Cubic yards = Cubic meters × 1.308
Cubic yards = Hectoliters × 0.1308
Bushel (2150.42 cubic inches) = Hectoliters × 2.838
Fluid drams (U.S. Pharmacopoeia, or U.S.P) = Cubic centimeters ÷ 3.69
Fluid ounces (U.S. Pharmacopoeia, or U.S.P) = Cubic centimeters ÷ 29.57

Fluid ounces. (U.S. Pharmacopoeia, or U.S.P) = Liters ×
33.82
Gallons (231 cubic inches) = Liters ÷ 3.785
Gallons (231 cubic inches) = Liters × 0.2642
Gallons (231 cubic inches) = Hectoliters × 26.42
Gallons (231 cubic inches) = Cubic meters × 264.2

Weight

Grains = Grams × 15.432
Dynes = Grams ÷ 981
Fluid ounces = Grams (water) ÷ 29.57
Ounces avoirdupois = Grams ÷ 28.35
Ounces avoirdupois = Kilograms × 35.27
Pounds = Kilograms × 2.2046
Pounds = Tonne (metric ton) × 2204.6
Tons (2000 lb) = Kilograms × 0.0011023
Tons (2000 lb) = Tonne (metric ton) × 1.1023

Unit Weight

Pounds per cubic inches = Grams per cubic centimeter ÷
27.68
Pounds per feet = Kilogram per meter × 0.672
Pounds per cubic feet − Grams per liter × 0.06243
Pounds per cubic feet = Kilogram per cu. meter × 0.06243
Pounds per horsepower = Kilogram per cheval × 2.235

Pressure

Pounds per square inches = Kilograms per sq. cm. × 14.223
Pounds per square inches = Atmospheres (international) ×
14.696
Feet of water (60°F) = Kilograms per sq. cm. × 32.843

Energy

Feet-pounds = Joule × 0.7376
Feet-pounds = Kilogram-meters × 7.233

Power

Horsepower = Cheval vapeur × 0.9863
Horsepower = Kilowatts × 1.341

Horsepower = Watts ÷ 746

Feet-pounds per second = Watts × 0.7373

Miscellaneous

Btu = Kilogram calorie × 3.968

Centimeters per second = Standard gravity (sea level 45° lat.) ÷ 980.665

Tons refrigeration = Frigories/hour (French) ÷ 3023.9

Water Factors

This section provides a guide for converting water factors based on a point of greatest density of 39.2°F.

Cubic inches = Ounces (of water) × 1.73

Cubic inches = Pounds (of water) × 26.68

Cubic inches = English gallons (Imperial) × 277.41

Cubic inches = U.S. gallons × 231.00

Cubic feet = Pounds (of water) × 0.01602

Cubic feet = English gallons (Imperial) × 0.1605

Cubic feet = U.S. gallons × 0.13368

Cubic feet = Tons (of water) × 32.04

Ounces = Cubic inches (of water) × 0.57798

English gallons = Cubic inches (of water) × 0.003607

English gallons = Cubic feet (of water) × 6.232

English gallons = Pounds (of water) × 0.0998

English gallons = U.S. gallons × 0.8327

English gallons = Tons (of water) × 199.6

U.S. gallons = Cubic inches (of water) × 0.004329

U.S. gallons = Cubic feet (of water) × 7.4805

U.S. gallons = English gallons (Imperial) × 1.201

U.S. gallons = Pounds (of water) × 0.1198

U.S. gallons = Tons (of water) × 239.6

U.S. gallons per minute = Miners inch (of water) × 8.976

Liters = English gallons (Imperial) × 4.546

Liters = U.S. gallons × 3.785

Pounds = Cubic feet (of water) × 62.425

Pounds = Cubic feet of ice × 57.2

Pounds = Cubic inches (of water) × 0.036124
Pounds = English gallons (Imperial) × 10.02
Pounds = U.S. gallons × 8.345
Tons = Cubic feet (of water) × 0.03121

Weights of Steel and Brass Bars

The following table provides a guide for finding the weights of steel and brass bars using a weight of a bar 1-foot long as a basis.

Size (inches)	Steel			Brass		
	lb	lb	lb	lb	lb	lb
$\frac{1}{16}$	0.0104	0.013	0.0115	0.0113	0.0144	0.0125
$\frac{1}{8}$	0.042	0.05	0.046	0.045	0.058	0.050
$\frac{3}{16}$	0.09	0.12	0.10	0.102	0.130	0.112
$\frac{1}{4}$	0.17	0.21	0.19	0.18	0.23	0.20
$\frac{5}{16}$	0.26	0.33	0.29	0.28	0.36	0.31
$\frac{3}{8}$	0.38	0.48	0.42	0.41	0.52	0.45
$\frac{7}{16}$	0.51	0.65	0.56	0.55	0.71	0.61
$\frac{1}{2}$	0.67	0.85	0.74	0.72	0.92	0.80
$\frac{9}{16}$	0.85	1.08	0.94	0.92	1.17	1.01
$\frac{5}{8}$	1.04	1.33	1.15	1.13	1.44	1.25
$\frac{11}{16}$	1.27	1.61	1.40	1.37	1.74	1.51
$\frac{3}{4}$	1.50	1.92	1.66	1.63	2.07	1.80
$\frac{13}{16}$	1.76	2.24	1.94	1.91	2.43	2.11
$\frac{7}{8}$	2.04	2.60	2.25	2.22	2.82	2.45
$\frac{15}{16}$	2.35	2.99	2.59	2.55	3.24	2.81
1	2.67	3.40	2.94	2.90	3.69	3.19
$1\frac{1}{16}$	3.01	3.84	3.32	3.27	4.16	3.61
$1\frac{1}{8}$	3.38	4.30	3.73	3.67	4.67	4.04
$1\frac{3}{16}$	3.77	4.80	4.16	4.08	5.20	4.51
$1\frac{1}{4}$	4.17	5.31	4.60	4.53	5.76	4.99
$1\frac{5}{16}$	4.60	5.86	5.07	4.99	6.35	5.50
$1\frac{3}{8}$	5.04	6.43	5.56	5.48	6.97	6.04
$1\frac{7}{16}$	5.52	7.03	6.08	5.99	7.62	6.60
$1\frac{1}{2}$	6.01	7.65	6.63	6.52	8.30	7.19

Steel—Weights cover hot worked steel of about 0.50 percent carbon. One cubic inch weighs 0.2833 lb. High-speed steel is 10 percent heavier.
Brass—One cubic inch weighs 0.3074 lb.
Actual weight of stock may be expected to vary somewhat from these figures because of variations in manufacturing processes.

Index